ABUBILAA

C000077115

THE
DIVINITY
OF
TIME
AND
COSMOLOGY

IRONHEART
PUBLISHING HOUSE

The purpose of this study is purely for an ambition of knowledge and enlightenment.

The contents of this book do not represent a universal view, because a universal view is not absolute and evident. No part of this book is an attempt at predicting or discerning future events or outcomes, nor is there an intent to impose views, doctrines, or ideologies.

We do acknowledge that there may be other intellectual views more accurate and authentic than ours, in which case, we encourage active participation and correspondence through channels enlisted at the end of the book.

Kindly approach this book and its contents with an open mind and an open heart. If a difference of opinion arises, we encourage the reader to do an impartial and objective study so as to present an alternate view by which both the reader and the author can benefit.

In the name of God Almighty, the Most Merciful, the Most Gracious, whose Divine Majesty perplexes the Hearts and Minds of many a great people.

In the Hearts of our Beloved Prophets, His Light guides them to guide us.

In the Hearts of the Enlightened Scholars of Islam, esteemed as they are, His Light guides the Prophets to guide them.

We are all but slaves and servants to His Greatness.

Like the thirsty in the desert, I yearn for every droplet of Knowledge, from the Holy Prophet of Allah Almighty, to whom I dedicate this book and the Enlightened Scholars of Islam to whom I dedicate this book.

Lastly...

To my Grandfather, Ramzan Harun.

May Allah Almighty shower you all with His Mercy and Blessings, and a special place in Jannat-ul-Firdaus.

Ameen

Time the Lonesome Road

It beareth the Journey of Life

To a realm beyond the Earth

The traveler must always strive

Is he not a fool then

Enchained is he to futile moments?

Unseen to him is the Hour and its portents

Enslaved is he to the Imposter's material thirst

When the World Beyond is far sweeter than the first

You may think yourself as insignificant a thing

Yet in the Heart resides the cosmos, its entirety within

See then the earth, a grain within a grain

See then reality like flesh, vessel, and vein

See then the Heavens unravel

Beauty like a blossoming flower

See then man eternal

In an eternity manifest of the Hereafter

On this journey, you walk alone

Time is your currency, Light is your guide

Your Heart with Knowledge and Wisdom, fortified

Venture forth, O' journeyman, upon this Lonesome Road

From the hardened earth, through the Sands of Time

Into the Cosmos beyond, and the Divine Light

There you will find your freedom

There you will find your Creator

INTRODUCTION
OF THE STUDENT AND THE STUDY
P13

OF RELIGION AND SCIENCE
P21

فَإِنَّهَا لَا تَعْمَى ٱلْأَبْصَرُ وَلَكِن تَعْمَى ٱلْقُلُوبُ ٱلَّتِى فِى ٱلصُّدُورِ

For indeed, they are not blinded by their eyes, but they
are blinded by their Hearts within their breasts
~Surah Al-Haj v.46~

OF
THE STUDENT
AND
THE STUDY

Biology, by definition is the study of Biological Life and Living Organisms. Sociology is the study of Human Societal Structure. By crude definition, Cosmology is widely defined as the study of the Universe. They are all considered various branches of Knowledge, and similarly, Epistemology is the branch of Knowledge which studies Knowledge itself. The most perplexing branch of Knowledge, is that of Chronology.

By linguistic definition, Chronology is not a study. It is, according to the English dictionary, 'the science that deals with measuring time by regular divisions and that assigns events to their proper dates.'

We, who will be delving deep into the fathoms of Time and Cosmology, will unveil new meanings to these words, more profound meanings of Depth and Divinity, to the branches of Cosmology and Chronology, both as the Islamic Sciences of Time and the Cosmos, not the 'clock' and the 'observable universe'. Together, our objectives will not be to repeat what secular science has broadcasted for so many centuries, such that their words have almost become true to our hearts and minds. No, our objectives will be to unravel the Divine Knowledge of Time and the Cosmos permeating from the essences of Islam and the Holy Qur'an.

Before we begin, we must take a moment to recognize the important difference between *'chance'* and *Divine Decree,* for both these terms are the reasons for the endless clash between the realms of sciences and the realms of Divine Knowledge.

The former lays the foundation of secular science within informative confines, while the latter enables the believer and knowledge-seeker to recognize the importance of science with limitations, and to affirm the overseeing and overruling sovereignty of Religious Knowledge over scientific information.

Science is not, simply put, a form of *Kufr*. The linguistic definition of *Kufr* is to 'cover', and a *'Kaafir'* is 'he who covers up'. In the context of religion, *Kufr* is the act of 'covering up the true message of Allah, *knowing that it is the truth'*. Therefore, science becomes *Kufr* when we overstep our boundaries and attempt to cover the Truth under shrouds of fancy words and equations for the sole purpose of validating our own doctrines and opinions. That is the Secular Scientist, even if he bears a Muslim Name and is abiding by every tenet of Islam.

Within limits, science constitutes the foundation of earthly knowledge, be it physics, chemistry, biology, medicine, arts, legislature, geology, whatever the branch of study. Science is the rational faculty by which Allah's Creation can explain and be explained in definitive and quantitative forms, because we as a creation have a definitive and quantitative response to all things logic and rational.

It is upon the righteous believer to acknowledge that Allah Almighty rules above all things rational and irrational (as per human perception). Creation adheres to His Command, finite and infinite. Even when formulating the minute and microanalysis of sciences, arithmetic and definitive descriptions of things, we must recognize that science is not the language of the Cosmos. It is a language developed by man for the sake of man's comprehension.

When we as humans define elements such as hydrogen and helium, or when we definitively state the age of the observable universe as approximately thirteen-billion years, we should know that this is *our* language meant to satisfy *our* logic and understanding formulated around *our* immediate environs. It is not the language of Allah Almighty, nor does it overrule or validate what He has revealed unto us through the Divine Scriptures. In essence, it is *His* Revelation that must confirm what human rationality conceives. If the Holy Qur'an does not confirm scientific theory, then science must be thrown in the bin. That being said, there is no reason for there not to be harmony, as we will elaborate further along.

The subjects of Chronology and Cosmology are not only complex in their own respects, but just as controversial, often resulting in vastly opinionated differences. There are two main reasons for the above mentioned intricacies, one being that aside from perception, both are beyond human sensory fields, by way of which human beings derive information, and hence, both can only be studied from a distance with a very finite and limited supply of information.

As such, both subjects must be handled with care and respect, because they are derivative of interpreted sciences, and human interpretations can be wrong. The Holy Qur'an is absolute, pure, and sacred, and one must ultimately refrain from trying to make an interpretation to fit one's own theories without validity. Theories are theories, hypotheses are hypotheses, and opinions are opinions. Neither become fact through the manipulation of Qur'anic scripts and translations. One must present, as we have throughout our studies, proper validation and irrefutable evidential weight, at least from those who have transcended a heightened level of knowledge and understanding, such as the great scholars of Islam, the Companions of the Holy Prophet, and all the Holy Prophets themselves.

Even in this respect, the subjects of Time and Cosmology, or any other ambiguous subject, must be studied with a modest and integral level of understanding both Science and Religion, and with utmost recognition that not only is the Word of Allah the *only* source of absolute Knowledge, but that only *He* has power to know *all* the interpretations of *all* the things He reveals to us and conceals from us of His Creation.

Let us now begin with understanding the limitations of Science, namely the oldest academic discipline and branch of science that deals with the 'naturalistics', 'arithmetics', and 'mechanics' of things.

The foundations of physical sciences, or Physics, are chiefly defined by the Three Laws of Thermodynamics.

The first law, the Law of Conservation, states that energy is always conserved in an isolated system, cannot be created nor destroyed. It can only be transferred, for example, turning on a switch to a bulb is the conversion of electrical energy into heat energy into luminous energy, and so on. The paradox of this law is that if energy cannot be created or destroyed, from whence does it exist in the first place?

The Second Law of Thermodynamics deals with the natural direction of energy processes. An example of this

17

law is that heat, on its own accord, will naturally flow from a hot environ to a cold environ. This again falls into the same contradiction as the first law. For energy to be transferred, not only must it exist in the first place, so too should the medium of transfer.

The Third Law of Thermodynamics is the Law of Entropy, and by far the most confounding of all three. Entropy is the measure of Order and Efficiency of energy in a system. A system is said to have reached maximum Entropy (disorder and chaos) when it has exhausted all its energy and order, becoming inefficient or 'chaotic'. In short, it is a quantitative measure of what could be defined as an 'absence of energy', similar to 'cold' or 'darkness'.

The Laws of Thermodynamics, hence the Laws of Physics, only hold true to human perception and with regards to the tangible compositions of the finite universe observed. By and large, they regulate the physical, chemical, and biological states of existence by adhering to a simple principle— *'everything that has a beginning, has an end'*.

The limitation is that these laws do not encompass the actuality of *intangible* concepts, such as those of Time, Knowledge, Wisdom, Light, and even as difficult as it may be to comprehend, the sublime existence of humanity beyond biological existence. In short, these Laws cannot explain, nor do they acknowledge *Aakhira*, or the actuality of the Hereafter, nor the actualities of the *Samawat* (Heaven), the *A'arsh* (Throne of Allah), *Jannah*

18

(Paradise), and *Jahannam* (Hell).

In summary, the Laws of Physics state that;

The universe will eventually 'die' in its own entropy, defined as the 'heat death' of the Universe, by way of exhausting all its fuel and energy.

The universe cannot have existed in any other form, aside from its physical form, before the Big Bang, and so will not exist in any other form after its death.

Both the above statements are given weight by an additional hypothesis called the Zero-Sum Universe, which foremost states that the universe 'emerged from nothing', and further states that the total equivalent energy of the universe is 'Zero' with positive energy balancing out with negative energy.

While human beings are in constant speculative debate over all these variating theories and hypotheses, we tend to forget one simple fact, which is a fact that can only be unveiled when one enters the realm of faith.

The Laws of Physics, hence the Faculties of Science, attempt to define the beginning and end of the observable universe without acknowledging the *Amr* (Command) of Allah Almighty and His attribute as *Al-Rabb*;

He (Allah) is the Doer of His Intent

[Surah Al-Buruj v.16]

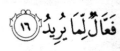

Which means that regardless of what the Physics say, Allah Almighty can intervene whenever He so Wills.

In the following chapters, *Insha'Allah*, we shall explore the limitations of science, the difference between truth and fiction within science, and the audacity and inevitable self-destruction of speculative and secular science before the absolute knowledge of Allah Almighty.

We will take another moment here to caution every Muslim pursuing this knowledge. Do not use science to prove the validity of the Qur'an and Islam. Do not use scientific discoveries to prove the Divinity of the Holy Qur'an, or to prove the existence of God. Allah Almighty and His Holy Scriptures do not need *us* to prove their existence. The pursuit of such notions is not only an insult to our own intellects, but an even graver insult to Allah Almighty and His Divinity.

He is the Creator of Worlds, because He is neither Cause nor Effect. He does not beget nor is he begotten.

OF
RELIGION
AND
SCIENCE

Within their definitive forms, the laws of physics *only* adhere to *physical* creation. Hence the term *physics*.

Of that which exists beyond the tangible, the physical, the Laws of Physics themselves disintegrate into paradoxical chaos and complexities.

It is without a doubt that certain elements existing to our perception cannot be quantified, whereby they do not adhere to Physics. By definition, even the essence of human existence cannot adhere to physics beyond its biological form and earthly attachment, hence the reason why secular science cannot explain our existence without the fabricated Darwinian theory of evolution.

When addressing the essence of human existence,

21

both the *Nafs* (Soul) and *Ru'h* (Spirit) supersede scientific rationale. Given this, the questions then posed are these;

Should science be the charter to which Religion must conform? Or should science be the branch which only studies the physicality of existence within the governance of Religion?

By all accounts, science defines the quantitative and perceptible outlines of tangibility, providing an explanation that adheres to the logical, rational, and plausible faculties of human perception. This faculty is defined as the 'Mind' and 'Intellect', or *Aql*.

The essence of intelligence is defined as a harmonious integration between the brain and the conscious, which is the wholesome integration between the body and the soul (the *Jism* and the *Nafs*). The reasoning is provided by the fact that intelligence is not only a neurological function, but an integration of 'sense' and 'actuation' as well. In technical terms, one can define it as the input of information, the processing of information, and the corresponding output of information.

For example, as a sensor, the eyes can be used to 'image' objects for the brain to process, and as the brain directs the eyes in vector, the eyes can also be used to 'see' as a responsive actuator. Similarly, the ears can receive information, and can also be directed to respond to sound. The mouth can be used to taste as

22

well as speak. The nose can be used to smell as well as draw breath. These sensory and actuation fields extend further into every component of the body, including internal organs which can react to positive and negative influences upon them.

In isolation, every bodily interaction with its surroundings is purely an informative interaction, and in several cases, the brain as the central processor can choose to interact with relevant inputs, and discard irrelevant ones. Just as well, every input and output remains informatory for as long as it is left unprocessed. It is only when a conscious interaction occurs that relevant information is processed and computed, thereafter allocated to a decisive reaction and stored to memory. Conscious intervention hence occurs on an individualistic level, the identity designated to the body, each one unique in its own creation. This is the *Nafs*. The Soul. The actuality of individual existence. This is the object to which a name or title is attached, to which character and personality are attached, and to which memory, thought, intuition, and inspiration are delivered and affiliated.

Across the spectrum, beyond logic and philosophy, there exists the other half of human existence, that which gives the human its humanity. The Life-force. The Spirit. The *Ru'h*. Logic and rationality lose their quantitative forms in the realm of the *Ru'h*, for here, only that which makes *sense* can be perceived, independent of plausibility. The integration between the conscious and the spirit, or the *Nafs* and *Ru'h*, gives the human a faculty of *Emotion*. This integration is housed within the 'Heart', or *Qalb*. This is not the physical organ, just as the *Mind* is not a physical organ. It is here that the existence of the human being bears significance on sublime and divine planes. It is here that the human can *feel* the positive elements of love, affection, sorrow, sympathy, empathy, and even the negative elements of despair, hate, envy, arrogance, ego, and anger.

The conflict of the *Nafs* is that it is ever in a turmoil to balance itself between the material and physical attachments of the *Jism*, and the spiritual attachments of the *Ru'h*. It is ever in turmoil between Logic and Emotion. Between Rationality and Sense. The test upon the *Nafs*, therefore, is in its ability to successfully bring harmony to the human being's existence under the regulated and divinely governed guidelines of Religion.

This harmony endows the human being with a unique ability, which is the ability to absorb Knowledge from *two* worlds. The modern world has struggled, and to a great part, is succeeding in elaborately airing that knowledge can only be acquired from one source; the institutions commission *by* the modern secular world. In doing so, it has widely misled humanity to believe

24

that the physical, material knowledge of this world is all that exists. What we know, from the Holy Qur'an and the teachings of Islam, is that Knowledge comes from *two* sources. Physical *and* Spiritual, and the point of intersection of both worlds of Knowledge is defined in Surah Kahf, in the event of Nabi Musa and Al-Khidr, as *Majma-ul-Bah'rayn* (where two oceans meet).

The Earthly knowledge of science, and the Spiritual knowledge of Religion.

What is the relevance, or rather *why* is it relevant to make this distinction?

There is a crude and detrimental, cancerous ideology arising within the society of the modern age, that science is *the* defining factor of our existence, hence the defining factor of religion. By extension, society is ever inclining towards science to explain the rationality of religion, and the consequence of this doctrine is that science *cannot* explain God. The resultant sees the emergence of atheistic belief, giving unwarranted precedence to secularism, where a vast populous then claims that it is 'spiritual' but not 'religious'. The question begged then is, *how do you find spirituality, if you cannot even adhere to an actuality of religion, let alone Islam?*

The entirety of human existence is as a result of cosmological creation, and by definition, the study of the cosmos is as a direct result of human perception. Without human observation, the science of cosmology

25

would only be glittering stars in the night sky with a 'flat earth' as the center of the observable universe. Undoubtedly, it is Religion that has provided the proof and extensive definition of the cosmos and human existence. The variance is that science corroborates the observable universe on defined scales, while religion defines the Cosmos in poetic and elegant actualities. In truth, the defined scales are irrelevant to the believer who understands that a 'defined life' on earth within physical and biological confines is only temporary. To the believer's *Heart,* that world beyond, defined in a poetic and eloquent sense, is far more enticing than this.

The Holy Qur'an explains this strange correlation between the limitation of materialism, and abundant gratification of the world beyond;

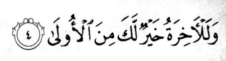

And surely the Hereafter is better for you, than the first (world).

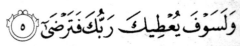

And your Lord will provide you, and you will be satisfied.

[Surah Ad-Dhuha v.4-5]

It follows, therefore, that the scientific mind,

Muslim or non-Muslim, must first come to terms with itself— Is the thirst for science for a benefit of this world, or the other? Because science itself can only define *this* world, hence science itself only exists on an informative plane. Science becomes Knowledge when the mind comprehends a Divinity to Creation, rather than a simplistic rationale.

The balance between *Jism* and *Ru'h*, or the *Majma-ul-Bah'rayn,* can only be attained when one is on a pursuit of True Knowledge, rather than a mere informative inquisition, and the origin of this journey must begin with a clear understanding that Knowledge is from a source Divine. Not the university.

The cosmos, and the study of cosmology, is truly a remarkable and mesmerizing realm. Oh, what a wonder upon mankind, to observe the Earth from the peaks of Saturn's moons, or to point at the Milky-Way from the vastness of the Andromeda. Oh, what a wonder upon mankind, to fully comprehend, with utmost satisfaction, the *'How'* and *'Why'* of the universe's existence, rather than just reading the *'What'* from a textbook full of numbers, words, and diagrams.

Saddening it is to see the fellow man, intrigued not with the existence of the cosmos, but with how much money he made today, or what new ambition he will pursue tomorrow.

In his *Mishkat-Al-Anwar* (The Niche of Light), Imam

27

AbuHamid Al-Ghazzali defines one of the 'defects' of Visual Sight as apprehending the large as small, or the significant as insignificant. For instance, the eye sees the Sun as the size of a small plate, and the stars as specks of dust on a black canvas, which he describes as 'silver-pieces scattered upon a carpet of azure'.

However, intelligence perceives the stars and the sun to be much larger, multitudes of the earth. To the human eye, the stars seem stationary within the moment, or the shadow cast by the sun as standing still, but intelligence perceives it to be moving constantly as it grows, lengthening with a motion of Time and Space (*Al-Furqaan v.45-46*). The stars speeding with every instant through the vastness of space (*Adh-Dhariyat v.47*).

It should be noted that what Al-Ghazzali, and others like him, define as Intelligence (*Aql*) is not only the scientific faculty of logic and rationality, but what we earlier described as the harmonious realm of *Majma-ul-Bah'rayn*, the convergence of earthly knowledge and spiritual knowledge. The faculty of 'intelligence' in the context of this study is defined as the 'Inner eye' by which *Nur* from Allah Almighty illuminates the path of he who is seeking *Majma-ul-Bah'rayn*. The Light of Intelligence is acquired through the process of converting Sensory Information into Comprehensible Knowledge through Thought and Contemplation as

28

well as through Spiritual Pursuit.

This analysis is made without the acknowledgment of error on the part of the observer by attributing the 'Knowledge received' to be the absolute truth from Allah Almighty, and by attributing Allah Almighty to be the Knower of absolute truth.

It follows, therefore, that while the intelligence of man does 'see', not everything it sees exists on the same plane. Knowledge is acquired as well as 'gifted' into intelligence by way of 'intuition' or 'inspiration'. The rational faculty of man attempts to sift between fact and what it perceives as 'fiction' by noting that one thing cannot be both with and without an origin. It cannot be existent and non-existent at the same time.

For example, granted the existence of 'black', it follows that color should exist, but 'color' does not include 'black'. Another example is that of Man and Beast, that man can become 'beast-like' in character, but a 'beast' cannot become 'man', thus refuting the doctrine of man having evolved from ape.

This example of 'evolution' falls under the category of 'Speculative Theorems', and requires a definitive 'spark' of truth, or an Absolute Light to illuminate the validity within the 'thought' for the Intelligence to 'see' the truth.

To secular science, this notion is highly displeasing and upsetting because it requires science to acknowledge

29

one thing; that above every textbook, only the Holy Qur'an can shed absolute light.

The verses of the Holy Qur'an, in relation to intelligence, have the value of Sunlight to the Sensory Eye. Without the Sun, the human eye could perceive very little, if anything. Now, again science will argue that without the sun, the human eye will eventually 'evolve' and 'adapt' to its sun-less surroundings, but arguments remain speculative until proven fact, and the challenge upon science is to present the human eye that can see in the dark without light, or any prosthetic assistance.

Here, we are understanding how the Divine Revelation of Allah, which is the absolute form of knowledge in our world, can act as the shining light illuminating the pathway of the believer by way of delivering 'sight' to the inner eye through intelligence.

This claim can be found in the Holy Qur'an itself;

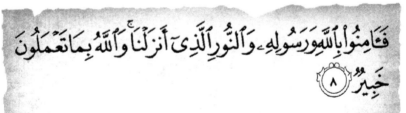

فَـَٔامِنُوا بِاللَّهِ وَرَسُولِهِۦ وَٱلنُّورِ ٱلَّذِىٓ أَنزَلۡنَا وَٱللَّهُ بِمَا تَعۡمَلُونَ خَبِيرٌ ۝

So believe in Allah and His Messenger, and the Light (the Knowledge of the Qur'an) which We have sent down; And Allah is All-Aware of what you do.

[Surah At-Taghabun v.8]

30

It follows, therefore, that the Holy Qur'an befits all human perception of Light unto intellectual sight, just as the Sun fulfills its role of illuminating our visual and sensory world. The Qur'an represents the Light of Allah as one of the branches from the blessed and pure Olive Tree (*Nur v.35*), the branch that delivers absolute knowledge to man's inner eye.

Allah Almighty says;

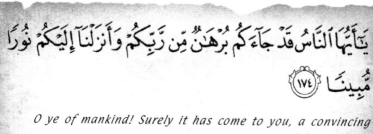

O ye of mankind! Surely it has come to you, a convincing (irrefutable) proof from your Lord, and We have sent down to you, a Light so clear.

[Surah Nisaa v.174]

PART ONE

THE
DIVINITY
OF
TIME
AND
DIMENSIONALITY

TIME ON EARTH

In every human tongue, the word 'earth' is almost unanimous, and in every instance, it not only denotes Earth as a planet, but as 'Home', the 'Beloved Ground', 'Sacred Dirt', because its every seam is interlinked with man's existence.

Aside from the First Man and Woman, Adam and Hawaa, every other human has only ever lived on the earth, is created of the earth itself, and shall return into the earth.

In Arabic, it is *'Ardh'*. In Hebrew, it is *'Adamah'*. In Danish, it is *'Jorden'* (pronounced 'yorde'). In Dutch, it is *'Aarde'*. In German, it is *'Erde'*. In Swahili, it is *'Ardhi'*. In Icelandic, it is *'Jord'* (yord). In Latin-derived languages such as Spanish, Italian, and Portuguese, it is

'Terra'. In Persian-derivative tongues, including Urdu, it is *'Zameen'* (also used to describe Land). In Greek *'Gaea'* denotes 'Mother Earth', as well as in the Indian subcontinental tongue, *'Prithvi'* or from old Sanskrit *'Prithvi Mata'* (Mother Earth), giving earth the same respect as one would give their Mother, as She who has borne us all.

Hence, every affiliation and attachment with the earth is, to human perception, *Earthly*.

Including Time.

Greek philosopher, Aristotle, said that 'Time is the unknown of the unknown', and so by extension, it is the unseen of the unseen. Yet, the actuality of mankind is that we are trapped and imprisoned in the 'Present Time'. We cannot step back into Time and alter an outcome, nor can we step forward in Time and redefine the present. This school of thought is guided by Logical Philosophy, and it only holds true so long as the perception of Time is done by a sensory analysis rather than internal intuition.

When we implement Intuitive Philosophy, we understand a different concept of Time. That Time is relative. Time is boundless. Time is independent, and Time *can* be traveled.

This conforms the Realm of *Spiritual Time*, as opposed to Earthly Time, or Physical and Biological Time, which we have grown accustomed to.

36

The distinction to be made is that Spiritual Time is boundless on Spiritual Planes, whereas Physical and Biological Time is bound to Physical Planes. As a rule of thumb, our dimension of earth binds us to our dimension of Time.

Time on Earth is defined by measurement, a concept which simply means— Time is what the clock reads.

In physics, Time is scalar, just like Length, Mass, and Charge, and is described as a Fundamental Quantity, which is a unit of measurement for a Base Quantity. It can be mathematically combined to measure other concepts such as Speed, Motion, and Time-dependent quantities. For instance, to measure the speed of a moving vehicle, one must account for the distance traveled and the Time taken to travel that distance. If the object has moved 100 kilometers in 1 hour, the Speed of the object is given as a derivative of distance over Time-taken, that is, 100 Kilometers per Hour.

Open a dictionary and index the word 'Time'.

One will find definitions such as *'the indefinite continued progress of existence and events in the past, present and future regarded as a whole'* (Oxford Dictionary), or *'the measured or measurable period during which an action, process or condition exists or continues'* (Merriam-Webster Dictionary), and the most materialistic definition *'the dimension of the physical universe that orders the sequence of events at a given place'* (McGraw-Hill Encyclopedia of

Science and Technology).

All the above, and several other definitions, attempt to answer the question 'What is Time?', and by using the word 'What', they seemingly only fail in trying to objectify an 'unobjectifiable' concept.

This physical concept of Time has now been widely accepted as the basis of Time. It is hence viewed as a Linear, indefinite progress of existence and events occurring in irreversible succession from the past to the present and the future. We use Time to quantify a sequence of events, to measure the duration of events, and also the gaps between events, while also quantifying a 'state of change' in material reality and conscious experience.

For eons, Time has been a curious subject, a vital component of study in Religion, Philosophy, and Science, but defining it in any particular aspect has consistently eluded every scholarly mind. The perception of Time being relative changes for every individual who tries to define it, literally or allegorically. Thus we have phrases such as 'Time stood still', 'It took a very long Time', 'I don't have the Time right now', and 'Things were different in my Time' among numerous other examples.

After lengthy studies and analyses, the overall conclusion has often been 'Time is an illusion' or 'Time cannot be understood' when even philosophical minds have failed to describe Time. This is because we always ask the wrong question— What is Time?

Our existence is by the Hands of our Creator. Hence, the knowledge of every aspect of our existence can only come from our Creator. Strange as it may seem, even Time itself has been explained by Allah Almighty.

A *Hadith Al-Quds* is Sacred Hadith, named so because unlike Prophetic Hadith which can be traced back to the Holy Prophet (peace and blessings be upon him), the authority of *Hadith Al-Quds* is traced back not to the Holy Prophet, but to Allah Almighty. It is hence the Word of Allah, not documented in the Qur'an as a verse, but revealed to the Holy Prophet through speech.

In a *Hadith Al-Quds*, Allah Almighty has told us that He is Time;

The Prophet (peace and blessings be upon him) said, Allah has said, *'The offspring of Adam abuse Ad-Dahr (Time). I am Ad-Dahr. In my hands are the night and the day.'* (Bukhari and Muslim)

Dahr (Eternal Time) is only one such word used to define Time.

In Arabic, Time can also be given as *Sa'at* (hour), *Zamaan* (age or generation), *Qarn* (age or epoch) *Waqt* (period), *'Asr* (Passage of Time), among many others

39

such as *Yaum* (day or period), *Shahr* (month), and *Sanah* (year).

The reason why the Arabic Language (not dialect) holds far greater value in Authentic Epistemology, is because the Holy Qur'an (Divine and absolute Knowledge) was revealed in Arabic and has been preserved so since the beginning.

All these translations of Time bear numerous terminologies and descriptions as well as numerous meanings, because in reality there is more to Time than literal translation and transcription, regardless of language, linguistic, and region.

From a human perspective, Time constrained to 'Earthly' attributes is finite (limited, quantified, bound), whereas Time beyond 'Earthly' attributes is infinite (eternal, everlasting, immortal).

In the context of the above Hadith, *Ad-Dahr* is chiefly described as Eternal Time, the existence of Allah Almighty being Everlasting, Immortal, and Boundless hence Time also being eternal and boundless.

The deeper we dive into this subject, the more we lose ourselves. It is not in our benefit to try to explain and define that which is clearly beyond our comprehension, rather it is to have Faith in the Unseen and Unknown. Instead of asking 'What is Time?', let us move on and try to understand how Time functions with regards to our existence.

Time can be allegorically defined as a Tree with infinite branches in an infinite state of existence with an Infinite Lord. Time exists in the Heavens. Time exists in various dimensions. Hence, Time can travel through the dimensions, as well as through the Heavens. We will explore each of these concepts as we progress through the pages, but we can deduce from the above that a branch of Time on earth must exist in its earthly form.

Through science, we know that the earth's orbit brings about night and light by which we count our days. We know that the earth's revolution gives us a measure of years. We know that the earth's tilt gives us a measure of seasons. Hence, as a means to measure, quantify and calculate our passing, and the passing of events and occurrences on earth, Time has been refined into finer particles of hours, minutes, seconds and milliseconds, and even further into nanoseconds.

The arrogance that man can have is that *we invented* the 'clock', and so we have 'understood' Time, whereas the clock is only a precise calculator of Time, just like the tape-measure is a calculator of Length, and we have come to believe that since this is the only evidence of Time we can 'see' and 'feel' and 'sense', then *this* is all that Time is.

For the materialist, Time is money. Every passing moment not earned, or converted into currency, is a

41

moment of loss. To emphasize this concept and imprint it into every mind, the modern age has deliberated the concept of payment through hourly wage and salary. Given that this is the state of the world, and a rather unwarranted necessity of survival, one must take a moment to think and understand whether it really should be the way of life.

The reality unveiled is that many do not see it any other way. Aside from a very small minority, most of us toil through school and college for the sole purpose of acquiring a decent paying job with a notable title. Thereafter, every ambition is determined by numbers linked to cents and dollars. The Car, the House, the Bank account, the Trust Fund, the Property, and Investment. Some have identified it as the '40-40-40 lifestyle', that is, working 40 hours a week for 40 years of your life, and retire with 40 percent.

Saddening it is to see even our fellow Muslims falling into the same traps. The Holy Prophet (peace and blessings be upon him) said, *'The Hour will not be established until the people compete with one another in constructing high buildings'* (Bukhari)

In its literal sense, one of the signs evident of the Last Hour is the exponential increment of tall buildings manifesting themselves throughout the globe, but in its symbolic sense, the Hadith emphasizes something far grimmer. It defines a crucial aspect of man's competitive nature. It was addressed to the Companions of the Prophet at the time, who would have hardly seen a tall building in their lifetimes, and were among those who competed in righteous deeds. It denotes the overwhelming fact that progress, in this modern age, is no longer measured by a man's religious and spiritual respect, integrity, righteousness, and intellect, rather it is chiefly measured by a societal outlook of influence, wealth, and power.

This need to rise the ranks in the eyes of society, drives every individual to amass as much wealth as he possibly can, with the strange notion in mind that all his troubles will wane with enough coin. The truth is that it will never be enough, and despite man constantly quoting 'Money cannot buy happiness', its pursuit endures. The resultant consequence is that mankind's most precious commodity, Time, is traded for currency to no end.

The Muslim who seeks to see with *both* eyes, must now reflect upon himself. How much Time do we spend in the pursuit of the mundane, against how much Time we spend in the remembrance of our Lord? Is one hour, out of a ten-hour workday, sufficient to increase *Imaan* and attain *Ihsaan?*

The Muslim who seeks to see with *both* eyes, must now reflect and realize that Time is *not* an illusion. Time is certainly *not* money. Time indeed waits for no man.

Time *can* be traveled by anyone who has the ability to ascend to a level where they can understand Time.

Memories take us back in Time, and dreams take us to the future.

What we do, and how we do, in the present, will determine the wonder of that journey and the marvel of its destination.

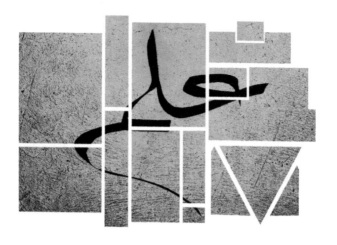

THE
RELEVANCE OF TIME

Perception is man's most precious tool for survival, hence many do not see beyond what they see, because 'seeing is believing'. A clouded perception will betray the Mind, but what if perception was a layer of delusion and illusion, instead of reality? Would perception portray an accurate actuality, or would it betray without even knowing it?

Regardless of opinion, the corpus of humankind relies on three fundamentals of existence. Namely Light, Time, and Knowledge, all taken for granted without a moment's thought. Here, at the precipice of the advent of the Final Hour, in the depths of the murky waters of the modern age, very few find the need to investigate their world, and fewer than few are able to penetrate the veils of the greatest deception to have emerged in the world. These few are of those who have embraced a reality beyond the material, by way of enabling their inner sights and perceptions to the actualities of Light, Time, and Knowledge. Not the kind propagated by modern secular institutions, but the kind divinely delivered through Scripture and Religion.

The world as we know it, has compounded itself into rules and dogmas, which are hence believed dogmatically, rather than perceived in a dynamic, meaningful, and comprehensive approach to Spiritual Reality. As crude as it may be to digest, such an outlook is barren of any essence, and is not endorsed by Islam on *any level*. Futile is every weak-minded attempt to 'Modernize Islam' as means of compromise, all for the sake of 'fitting in' to the modern godless age.

This compromising stance is regarded purely as a Formalistic outlook, and is intolerable under the umbrella of Islam.

The Naturalistic outlook, one without a proper religious guide, is always confined to Physical Reality, and even in the case of the Muslim, it leads and misleads to a materialistic approach to life and its enduring test. It is not, however, entirely condemned, for it adheres most to mankind's most primitive logical instinct. Hence the human urge to solve every one of life's problems from a practical perspective.

46

Not that physical reality is non-existent from a Religious outlook, but that Religion expounds by narrating that there is more beyond the physical confine, providing us with a greater, more coherent and illuminating view of life, thereby giving humanity a significance larger than physical and biological life and death.

However, physical reality is exponentially gaining greater attention and confinement as the entirety of the human race conforms to the growth of science instead of adhering to the Divinity of Religion.

The Religious view of reality, attested by the Heavenly Verses of the Holy Qur'an, has an Integralist outlook, wherein the spiritual view holds primacy over the physical view, constantly elaborated by the concept of life after death. It does not only provide all the formal grounds for righteous life (*Islam*), but cultivates an emotional attachment to divinity (*Imaan*), by way of dedicating the entirety of one's existence to God (*Ihsaan*), essentially transcending life from planar existence to spiritual destiny.

This analysis is given weight by the principle that the essence of human existence is the *Ru'h* (Spirit) and *Nafs* (Soul) both of which belong and originate from a Transcendental Realm of Existence. By definition, the human being is a spiritual being, hence should conform to spirituality for succession to higher planes.

If the spiritual realm exists on a sublime plane of divinity, it cannot abide by any *physical* concepts of motion, vector, and displacement, and so logically it must conform to rules of governance beyond the scientific rules of physics. These rules of governance must adhere to a concept of Time, without which neither vector nor displacement can exist.

Herein do we find the dire need to understand Time, because in our physical world we may be able to perceive Time as a measurement of progression, but Time in its origin cannot be quantified. Its relevance must be observed through internal intuition and insight, not as a quantifiable object, but as a subliminal plane which governs the vector of every one of Allah's Command as it transcends from His Throne to its designated point, manifesting into Creation.

This is a vital tool in understanding the 'Beginning and End', and that which preempts the beginning and concludes the ending. To surmise, before delving deeper, the Command issued falls into the stream of Time, its beginning determined from the point of origin, the *A'arsh* (Throne of Allah), its evolution transcending the stratas, arriving at its destination in the form originally decreed by the Almighty. The entirety of this process follows along the stream of Time, and every point of its journey is documented as the 'Knowledge of its Existence'. Knowledge is acknowledged when existence is actualized. Before its creation, Knowledge is concealed, hence Knowledge is revealed when the Creator wishes to reveal it, afore its emergence into reality, or at its exact moment of manifestation. Hence, all Knowledge is Revelation, and all Revelation is revealed in accordance to a motion of Time.

The concept of Time, therefore, is that it cannot be understood without interpreting the Knowledge of the Creation in study. Vice versa, the Knowledge of Creation cannot be comprehended without respectfully linking every advent to its point in Time.

As an example, we can analyze the formation of Islam into the defined structure of religion it is today. The formal, informative mind would classify the origin of Islam from the point in Time when the First Revelation and Command of *Iqra,* 'Read' (*Al-A'laq v. 1*) descended upon Nabi Muhammad (peace and blessings be upon him) in Cave Hira, on Jabal Nur, on the outskirts of Makkah.

However, from a knowledgeable perspective, the definitive Structure of Religion far supersedes Nabi Muhammad's Time. Its origin begins with the Command of Allah, as 'Be', and so the command to Create Islamic Structure begins from His Throne. Subject to the motion of Time, the structure of Islamic Laws and Guidelines actually take their infantile creations by the Prophethood of Nabi Ibrahim, and evolve, subject to

Time, through his progeny. The covenant formed with Ibrahim and his progeny descended from both his sons, the Israelites and Arabs, can be found in the Qur'an;

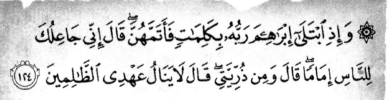

۞ وَإِذِ ٱبْتَلَىٰٓ إِبْرَٰهِـۧمَ رَبُّهُۥ بِكَلِمَٰتٍ فَأَتَمَّهُنَّ قَالَ إِنِّى جَاعِلُكَ لِلنَّاسِ إِمَامًا قَالَ وَمِن ذُرِّيَّتِى قَالَ لَا يَنَالُ عَهْدِى ٱلظَّٰلِمِينَ ﴿١٢٤﴾

And when his Lord tried Ibrahim with His Word (His Divine Testament), and he (Ibrahim) fulfilled them, He (Allah) said, 'Verily! I have appointed thee a leader upon mankind', he (Ibrahim) said, 'And from my progeny?', He (Allah) said, 'My covenant does not extend to the wrongdoers.'

[Surah Baqarah v.124]

Thus do we see the origination of a Structured Islam, as it transcended through the Israelites and the Arabs, converging to its final destination as the 'Complete Deen' revealed to Nabi Muhammad.

... ٱلْيَوْمَ أَكْمَلْتُ لَكُمْ دِينَكُمْ وَأَتْمَمْتُ عَلَيْكُمْ نِعْمَتِى وَرَضِيتُ لَكُمُ ٱلْإِسْلَٰمَ دِينًا فَمَنِ ٱضْطُرَّ فِى مَخْمَصَةٍ غَيْرَ مُتَجَانِفٍ لِّإِثْمٍ فَإِنَّ ٱللَّهَ غَفُورٌ رَّحِيمٌ ﴿٣﴾

...This day I have perfected for you your religion, and I have fulfilled upon you My favor, and I have approved for you the Religion of Islam...

[Surah Ma'idah v.3]

Islam by its definition is 'Submission', and from the beginning of humanity, mankind has been created for the sole purpose of Submission. Therefore, the precedence for an Islamic Structure, Religion, and Way of Life, took place with Nabi Adam (peace be upon him) the first Vicegerent on Earth, and was perfected by Nabi Muhammad (peace and blessings be upon him), the Seal of the Prophets. In addition, the Religion of the Israelites was also *Islam,* albeit under a different name and language. What is today regarded as 'Judaism', was historically termed by the conquering Romans during the 1st-century BC and 1st-century AD, when they renamed Judah (Hebrew, *Yehudah*) to Judea, and referred to the Israelites as Judeans, who later became Jews, and their religion was named Judaism. Strangely so, a majority of modern Jews do not know this.

Only by understanding the element of Time passing through the ages, from the epoch of Nabi Ibrahim to the final age of Islam, can we then comprehend the reality of events compounding our history.

Similarly, we cannot decipher our modern age by adhering to stationary events in spontaneous intervals or occurrences. Understanding the concept of Time beyond physical confines enables us to understand the Command of Allah and its manifestation into His Creation.

Therefore, when Allah Decrees the End of Mankind,

we may be able to deduce the aspect of 'What is happening?' or 'What may happen' by attributing a stationary perception of Time, but that only limits our understanding to the physical state of things. Beyond that we must ask 'How and Why is the current state of the world the way it is?' and with a relative understanding of Time, we transcend the physical confines of Information into the spiritual realm of Knowledge.

What is our understanding of Time?

We perceive Time to be linear, moving steadily, measurably and quantitatively in one direction— ahead. Throughout history, man has fantasied traveling back in Time to alter the past to better the present, or go ahead in the future and discover what awaits.

The natural incline of man is 'to know', and the urge to travel through Time arises from a state of internal restlessness and agitation. The 'need to know'. Hastiness arises and increases when we seek to know without having to make the *effort* to know, and the demand for immediate results preempts an insatiable need to navigate through shorter paths to the answers we seek. Hence the urge to travel ahead, in order to perceive the forthcoming unknown of our existence. For the large part, this 'need to know' conforms to the core of material acquisition, or the loss of material possession. By default, he who has no attachment to material possession, has nothing to lose, and therefore nothing to fear.

He who adheres to fear, fears all. Including the decay of Time. Within this narrow perception, Time appears linear and unidirectional, and in order to understand it, we hastily seek a quantitative and rational definition of Time through science.

Science, in itself, is not a fabrication. Science, when understood correctly, is in fact true knowledge, defining a comprehension of the physical and biological world which is also governed by a physical and biological Passage of Time. If Time exists in our physical world, then Time *can* also be understood through science.

The paradox is that because of the limitation of Man, so too is there a limitation on science.

Why is this relevant?

Within our own lives, existence is complex, and this creates a complexity of Time, because on the surface, without any thoughtful effort, we take Time as a perception. This perception creates the possibilities of illusions and delusions, which only add to the complexity of life, and this, at its core, is the conundrum we face when attempting to unravel the reality of the modern age.

A profound example can be drawn from the tale of the Companions of the Cave (*Surah Kahf*) and a detailed analysis has been made in *The Three Questions*.

Their perception of reality eluded them when they rose, even though hundreds of years worth of events had occurred during their isolation. As an example, when they questioned each other, their perception of Time had only endured a day or half a day. By that understanding, the Companions did not pay heed to their surroundings with the assumption that nothing significant would have transpired within that short span of Time. They could not comprehend the length of several hundred years of sleep, because every encounter with spiritual and empyreal transports, causes us to lose sense of Biological Time.

The 'here and now' concept we live by. The 'moment' we are entrapped in.

Only those who have attained such level of spirituality, as explained in Surah Al-'Asr (see the following chapter), can truly transcend to a higher plane of understanding.

TIME DIVINE

As we begin to unravel a profound understanding, we see that Time, in essence, is complex and multidimensional. It is beyond our linear and quantitative perception. This is attested by its existence beyond the definitive existence of man.

This point is crucial, because it outlines the thin line between secular science's propagation of man's evolution over millions of years, as opposed to religion's concise definition of man's finite and biological existence. In other words, the theory of biological evolution advocates the origin of man as a derivation of a pre-existing species (ape), implying that man's nature is subject to a continuum of evolution— from man to something else perhaps (Ape to Human to Transhuman to Posthuman).

By extension, this insinuates that biological Time is everlasting. If biological Time were to cease, so too would the existence of man.

The secular mind is appeased to abide by this ideology simply because the secular mind cannot comprehend the concept of death, and in a nutshell, this ideology arises from an instinctive fear of death as a physical affliction on the biological body.

In contrast, religion teaches us that biological confines are but a phase in the reality of life. Mankind was not created immortal in biological constraints, but the immortality of man's 'self', his 'identity beyond the body' transcends into eternal realms after the phase of biological life is complete. The transcendence is governed by the rule of finite biological existence, and herein lies the concept of *Aakhira*.

Its relevance to Time, therefore, is that Time in its true form is independent of man.

Time existed when we did not.

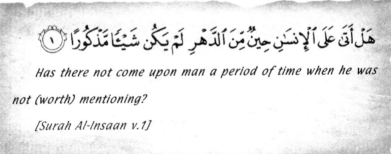

Has there not come upon man a period of time when he was not (worth) mentioning?

[Surah Al-Insaan v.1]

'Unworthy of mention' symbolizes the nihility of man, whereby the Almighty's intent of creating man had not even been overted to the Angels. This state of unworthiness is given by the phase of existence before Allah Almighty made His declaration to the inhabitants of the Heavens, that He would be placing a *being* on earth as a vicegerent (*Baqarah v.30*). The reality of man had not yet taken corporeal form. This phase also extends in duration up to the point when Allah Almighty breathed life into Adam, thereby declaring the 'finality' of His original decree of creating man. The Passage of Time is affirmed by the duration between His Command of declaring His intent, following into the manifestation of Creating man as a Creation, preceding the actuality of man's existence.

It follows, therefore, that the scientific theory of biological evolution is negated at its core. Meaning, the existence of man was *not* as a result of the evolution of a species, rather from a point of origin being one man, the first man and Vicegerent on earth, Adam (peace be upon him).

By definition, anything that has a beginning must have an end. Now we come to the concept of *Aakhira*. It must be understood that *Aakhira* does not denote the absolute End of mankind, rather it denotes the End of the Biological Phase in the existence of mankind. It, therefore, implies that the era of man 'in flesh and bone' will come to its finite end. This again negates the scientific theory of evolution that man would transcend into another species on earth due to an increment of its intellect, widely propagated as 'Posthuman' (refer to *The Abyss of the New World*)

The concept of *Aakhira* also affirms that Time is independent of man.

Time will exist when we will not.

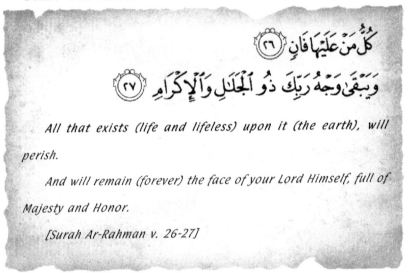

All that exists (life and lifeless) upon it (the earth), will perish.

And will remain (forever) the face of your Lord Himself, full of Majesty and Honor.

[Surah Ar-Rahman v. 26-27]

The final phrase *'will remain forever the face of your Lord'*, implies the immortality of God Almighty, His eternal existence, by which it is irrefutably implied that so too will Time continue to exist. In other words, for as long as Allah Almighty exists, so too will Time. This is given by His direct declaration in *Hadith Al-Quds* (Allah's direct speech); *'I am Time.'*

By the Passage of Time during man's creation, prior to man's predestination on earth, it can therefore be understood, irrefutably, that Time also exists in the Heavens, beyond quantitative and vector form.

In various verses of the Holy Qur'an, Allah Almighty speaks of Time-related events and incidences, describing a period of three hundred years, or one year, or thousands of years, half a day or even an hour. Here is where human comprehension falls into chaos, because it is difficult to perceive something intangible existing beyond the spherical, three dimensions of the earth. The human brain cannot comprehend anything it cannot quantify, and even though quantitativeness is possible in certain aspects, their significance eludes us.

Time defined in the Holy Qur'an, can become frustrating and perplexing to fully understand, if but we limit our understanding. If taken only in their *literal* meanings, or translations, the verses remain just so, albeit filled with awe. One would *count* 'a day to a thousand years' in a literal sense, when it only pertains to 'man's reckoning and perception', and does not necessarily sum up to a thousand earthly years (*Al-Hajj v.47*) or fifty thousand earthly years (*As-Sajdah v.5*). One would hasten to decipher *'a day like a year, a day like a month and a day like a week'* (Bukhari), in literal, quantitative measures, without fully understanding the sublimity of Time.

How do we relate to fifty-thousand years, when we cannot even relate to yesterday in its entirety? As we coerce our memory into quantifying the last 24 earthly hours of our life, conclusively, we can only quantify a mere fraction of it.

This does not mean that comprehension is beyond us. It simply means that comprehension does not necessarily require rationalization, or in other words, things do not have to be *rational* for them to make *sense*.

The foremost rule to be applied is that we cannot relate to anything divine while basing calculations and computations on the foundations of the physical world. We have to cross our physical realm of Knowledge, into a much higher, spiritual plane of Knowledge, in order to fully comprehend the definition of Time as is described in Divine sense.

One such example from the Qur'an is;

إِنَّ رَبَّكُمُ ٱللَّهُ ٱلَّذِى خَلَقَ ٱلسَّمَـٰوَٰتِ وَٱلْأَرْضَ فِى سِتَّةِ أَيَّامٍ ثُمَّ ٱسْتَوَىٰ عَلَى ٱلْعَرْشِ يُدَبِّرُ ٱلْأَمْرَ مَا مِن شَفِيعٍ إِلَّا مِنۢ بَعْدِ إِذْنِهِۦ ذَٰلِكُمُ ٱللَّهُ رَبُّكُمْ فَٱعْبُدُوهُ أَفَلَا تَذَكَّرُونَ ٣

'Verily your Lord is Allah, the One who created the Heavens and the Earth in six days, then he positioned Himself on the Throne (of authority), from whence He governs all. None can intercede (nothing can happen) without His permission. Such is Allah, your Lord. So worship Him. Will you not then take heed?'

[Surah Yunus v.3]

When Allah Almighty is speaking of 'Six Days', He is not giving us a 'count of time' by which He created the Cosmos. He is giving us the 'Stages of Time' that underwent this creation.

Quantifiable from science, we can somewhat deduce that the observable universe began from a singularity as a point of origin, thereby undergoing an expanse of formation. The splitting of a particle, the dense gaseous state forming atoms and molecules, shaping themselves into stars and galaxies, and so on and so forth. The rationalization of this theorem is that it only defines 'what happened', without bearing the capacity to accurately define 'how it happened' or 'why it happened'.

In contrast, religion fully explains the 'how' and 'why' by enabling the human mind to expand its ability to *think* and *contemplate*.

A concise example of such thought is as follows;

Creation begins with the Creator and His Command. His Command issued as 'Be' and from the singular point, His Creation begins to take form. Here the Passage of Time carries out His Command into His desired manifestation, and the 'six days' define the Six Epochs or Ages of each stage of manifestation. 'Six' as a number holds no significance to a 'count of Time', compared to 'Six' as a measure of 'how many ages of Time' the process underwent. Within this, we see a dimensionality of Time, or multiplicity of Time, where the various events manifest *Sequentially*, *Cyclically*, in *Parallel*, or *Independent* of each other.

This represents chaos for the rational mind, and harmony for the enlightened one. Time becomes complex if we try to define it as an object, rather than understand its essence. In its own respect, science is the faculty of knowledge which enables us to quantify the tangibility of our existence. Time being intangible cannot be understood through science, and so the answers we seek can only be found in the Qur'an;

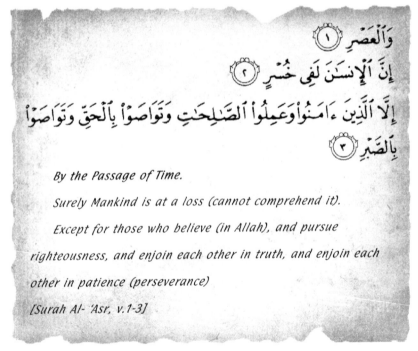

By the Passage of Time.

Surely Mankind is at a loss (cannot comprehend it).

Except for those who believe (in Allah), and pursue righteousness, and enjoin each other in truth, and enjoin each other in patience (perseverance)

[Surah Al- 'Asr, v.1-3]

The phenomenal knowledge contained within these three verses alone can enable us to write several volumes explaining the depth of this Chapter in the

Qur'an, and even then we would not have exhausted the knowledge contained, if but a little. Indeed Imam Shafi'i, one of the great Scholars of Islam, asserted that had Allah Almighty revealed *only this* Surah, it would have sufficed as guidance for humanity. In its simplest form of understanding, the Surah denotes that 'True comprehension of Time eludes man's capabilities within physical and biological confines'. Religion affirms that spirituality can only be attained through Faith, Righteousness, and Patience, and only spiritually can Time be understood.

The implication of all this is that Time is a Divine element.

Through Divine governance, Allah Almighty has enabled us to perceive Time in our biological confines, not the essence of Time, but a *measure* of Time.

وَجَعَلْنَا ٱلَّيْلَ وَٱلنَّهَارَ ءَايَتَيْنِ فَمَحَوْنَآ ءَايَةَ ٱلَّيْلِ وَجَعَلْنَآ ءَايَةَ ٱلنَّهَارِ مُبْصِرَةً لِّتَبْتَغُواْ فَضْلًا مِّن رَّبِّكُمْ وَلِتَعْلَمُواْ عَدَدَ ٱلسِّنِينَ وَٱلْحِسَابَ وَكُلَّ شَىْءٍ فَصَّلْنَهُ تَفْصِيلًا ۝

And We have made the night and day as two Signs. Then we obscured the sign of the night, and We made the sign of day visible, so that you may seek bounty from your Lord and know the number of years and the count (of time). And we have explained everything in detail.

[Surah AL-Isra v.12]

63

The cosmological positioning and displacement of the Sun, Moon, and Earth, as vector elements on a Space-Time continuum, conform to quantitative and uniformly cyclical motions, by way of which we, grounded to the spherical dimensions of the earth (conforming to the cyclical motion of Time and Displacement) can quantify the motion of Time respective to our physical and biological progression through life.

The manner in which Time conforms to the essence of existence between the Command of Allah and the Creation of Allah is truly mesmerizing in the sense that everything we can observably attest to, occurs with such precision and accuracy, without a particle's worth of deviation. Even the slightest of cosmological deviations in Time can result in our utter annihilation. As a consequence, we, the servants and slaves of Allah Almighty, must shed our arrogant selves of becoming 'too knowledgeable' on our own merits, and adhere to supplicating before Him in gratitude and worship.

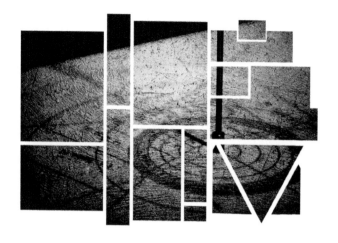

FLUIDITY
AND
RELATIVITY

Arithmetically, the shortest distance between two points is a straight line, regardless of how far apart they are. Spiritually, the shortest distance is but a thought away, but the reality of biological life represents something far more complex. The distance may be a straight line, but the journey never is. The path wavers and zigzags, twisting and turning, riddled with obstacles and trials.

Yet the journeyman presses forward.

Consider the following hypothesis.

A man intends to travel a distance from one point to another. On foot, the journey would cost him several hours. Given that he begins the journey at the mark of 00:00 hours on the clock and arrives at 12:00 hours, the entirety of the journey would cost him 12 hours of his life to complete.

If he takes a car instead, the journey would only cost him an hour. He begins the journey at 00:00 hours on the clock and arrives at 01:00 hours. He has arrived at his destination in an hour.

A surface analysis shows that he traveled quicker in space, from one point to another, but a deeper analysis shows that he, in fact, traveled faster through Time. In essence, he traversed into the future quicker by car than on foot. The defining factor here is that he made a decision between taking a slower journey as compared to a hasty one. The former would have allowed him to embrace the reality of the world better, whereas the latter only endowed him with reaching his destination sooner.

Given then, that it is the journey that matters, not the destination, here lies the question— In which of his decisions would the journey have been more profound and valuable?

Very like the tale of the Bedouin and the modern-man. The modern-man asked the Bedouin, 'How long does it take you to cross the Desert?'

The Bedouin replied, 'On my camel, two or three weeks.'

So the modern-man laughed, and said, 'In my airplane, I can cross the Desert in half a day.'

And the Bedouin smiled and asked, 'What would you do with all the time you save?'

He was smiling because he knew what the modern-man's endeavors were while he would be in the desert, on his camel, looking up at the stars, saying, 'Glory be to Allah, what a marvel is His Creation of the Heavens and the Earth.'

The substance or value given to Time spent, individually and holistically, defines the relativity of Time. It is not given by the lengthening or shortening of Time within quantitative forms, but by its intrinsic value from every observer's perspective.

In 1905, while commuting home from his patent office, Albert Einstein observed the manner in which he moved, with relation to a city clock. Given his speed of travel, he argued that the clock-hand moved directly in proportion to his own motion, he being the observer, and he hypothesized that were he to move faster, the clock had would appear to move slower. Given the fact that he could see the clock-hand by way of light bouncing off the clock and into his eyes, he then hypothesized that if he were moving at the speed of light, the clock-hand would appear to be motionless to him, the observer and traveler. By this observation, he concluded that Time moved differently for every observer. His conclusive studies and calculations introduced a new framework into Theoretical Physics, proposing new ideas for understanding Space and Time.

In his Theory of Special Relativity, he proposed that Space and Time were woven into a single continuum known as Space-Time, that is, 'events that occur at the same Time for one observer in one position could occur at different times for another'. In quick summary, the motion of Space abides by the motion of Time, both interlaced. Theoretically, if one were to change his physical coordinates in Space, so too would he alter his perception of the motion of Time, throughout the observable universe.

Here, the reader and researcher must adhere to an underlining fact that Einstein and Science did not introduce the concepts of Time by way of scientific discovery and breakthrough. They only rationalized pre-existing concepts into modern dialect and compositing, by way of which a mechanical measurement of Time could be understood with regards to modern perspective and scientific advancements. The theory of relativity is not 'new' to human knowledge, and while it remains a theory within scientific confines, it is a definitive fact from Qur'anic revelation. Meaning that the knowledge of Time Dilation and Relativity was not newly discovered, only that the information and its medium of relay was newly introduced with some scientific rationality.

In essence, the knowledge pertaining Time, and its evolution, was revealed to the Believers many centuries prior, and by extension, was also revealed in previous scriptures. Historically, even civilizations such as ancient Mayans, Egyptians, Chinese, Babylonians and Sumerians, also had a certain perception of Time, its motion through the cosmos, and its relevance pertaining the biological limitations of man.

Relativity simply classifies Time as 'not being linear and unidirectional'. It gives Time a certain fluidity and flexibility by way of which it can relate not only to the earth and the cosmos as a whole, but within the various dimensions of the earth and the observable universe.

The concept of relativity can be found in the Qur'an, and while the following verses are subject to interpretation, we urge the reader and researcher to abide by the profound thought that only Allah Almighty truly knows best their correct interpretations, and by extension, only He can attest to the accuracy of our explanations.

The Holy Qur'an says;

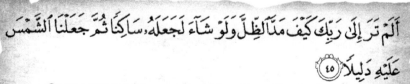

Do you not see how your Lord lengthens the shadow, and had He willed, He would have made it stationary. Then We made the sun for it (the object casting the shadow) an indicator.

69

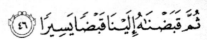

Then We withdraw it (the shadow) towards Us, gradually.

[Surah Al-Furqaan v.45-46]

While the modern clock provides us with a mechanical accuracy of counting Time, shadows governed by the motion of the sun have always been the natural way of perceiving the motion of Time. Objects such as sundials are less advanced by modern technological standards, but were near accurate instruments of measuring the Passage of Time in their respective ages, and hardly have they ever deceived humanity from determining the time of day.

The lengthening and withdrawal of the shadow, in the verses above, elaborate a Relativity of Time as 'something which is in motion', and had Allah Almighty willed, surely he would have left it stationary, by way of making the earth a stationary planet, instead of a rotating one. This is given by the implication of the Sun as an 'indicator' as it passes through the day, lengthening the shadow and gradually shortening it as it reaches its apex. Thereafter the shadow gradually lengthens again as the Sun completes its cycle, or rather as the earth continues its rotational motion into night.

The next verse describes the motion of the earth in reference to the shadow and its relation to Time as it

70

wanes through night and day.

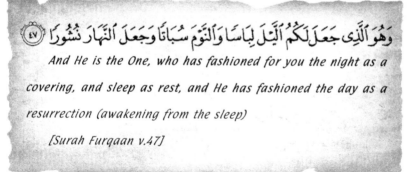

And He is the One, who has fashioned for you the night as a covering, and sleep as rest, and He has fashioned the day as a resurrection (awakening from the sleep)

[Surah Furqaan v.47]

The relativity of sequential Time in the above verse as 'day into night' is described as;

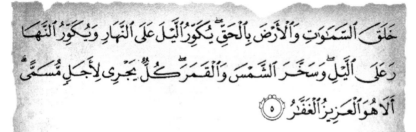

He Created the Heavens and the Earth in truth; He wraps the night over the day and wraps the day over the night; and He subjected the sun and the moon, each one on its (orbital) course for a term appointed; unquestionably, He is the Almighty, the Oft-forgiving.

[Surah Zumar v.5]

Relativity here is described in the process by which night overcomes day, and vice-versa, like the covering of one over another, or an 'overlap'. The gradual motion of one over another gives the day its dusk and dawn, two periods of Time quite variant from each other and variant

71

from the rest of the day and night. These interludes are also variant from the observer's coordinates on the earth itself. At the equator, day and night are equivalent to each other, so are dawn and dusk. In the northern or southern hemisphere, depending on the season, the balance of day and night varies drastically, and so do the interludes of dawn and dusk.

The seasonal expansions and contractions of Time are given by;

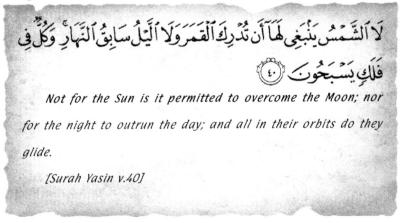

Not for the Sun is it permitted to overcome the Moon; nor for the night to outrun the day; and all in their orbits do they glide.

[Surah Yasin v.40]

More knowledge is derived from the motions of the Sun and the Moon. The moon runs its course in equivalent phases, each one relative to the other, in *Cyclical* Time variant from *Sequential* Time, vastly variant from *Paradoxical* Time and *Inter-dimensional* Time.

With relevance to the earth, the Cyclical Time within the seasons is also relative to each other, neither conflicting, but always in harmony.

If the day is longer in one part of the earth, the night

is equivalently longer in another part, maintaining a holistic balance in the motion of earth's mechanical Time.

As a whole, the sun is also in motion, coercing through the galaxy, taking the entire Solar system with it by imposing its gravitational pull on the Space-Time fabric upon which it rests, and also in its appointed term (Direction and Speed).

In physics, this is known as a Galactic Year, or Cosmic Year, and it is the duration taken by the Solar system to complete one full revolution around the center of the Milky-Way. The Speed at which the revolution is taking place is scientifically calculated as approximately 828,000 kilometers per Hour, based on earthly Time. To put it into relative perspective, it is a speed by which an object could circumnavigate the earth in approximately 2 minutes and 54 seconds.

Within all these varied intervals of Time, there exist transitional moments of one moving over to another, by similar way of the 'cover' described in the Verse above.

We, on earth, can call out the dates of the equinox, the fourteenth of every lunar month as the full moon, the 6am Sunrise and 6pm Sunset, even calculate the vast distances of space or the speeds at which celestial bodies are coercing through Space, but at the most fundamental level, all these remain as mechanical measurements of Time, not Time itself.

To many, the figures and numbers mean little in daily aspects, incomprehensible and in some cases, insignificant. They are taken for granted as norms that have endured for eons. They provide a brief flare of exhilaration and a satisfaction of a momentary thrill, after which they bear no meaning to the average man only concerned with the progress of life, hence the average man only concerned with material acquisition.

To the believer, and Knowledge-seeker, however, the Beauty of Time is the reddening of leaves in autumn, or the crescent moon at the beginning of every month, the chirping of birds at dawn, and the amber glow of dusk.

The former connotations secularize Time, while the above enlighten us to the essence and Divinity of Time, giving us a deeper understanding of the Fluidity and Relativity of Time.

TIME AND EVENTS

There is an element to Time, whereby it appears to move faster or slower, condensing or expanding with relevance to the observer. It appears to drag on forever when the mind finds itself wandering, and seems to pass quickly when the mind is preoccupied.

An example of this phenomenon occurs as the human mind develops. As a child, Time seems to stretch into boredom, because as active as a child's mind is, it lacks in acquiring activities to fill. The voids left with inactivity lengthen or shorten, depending on what the mind is demanding.

In contrast, an adult mind perceives Time as too little, or too short, filled with a constant bombardment of activities. In some instances the activities become so exhausting, even sleep seems insufficient by a measure of Time. Hours become days, days become years, all fleeting by.

The perception also differs between a mind developed in urban lifestyles versus one developing in the countryside. It differs between one who works with physical strain as opposed to one who works with mental strain. The academic, for whom hours can endure endlessly, perceives the Passage of Time differently than the soldier in the battlefield, whose pocket of Time left is but a bullet away.

This variance of Time lengthening or shortening is not governed individually, nor is it governed psychologically as widely assumed. The Passage of Time is responsible for one thing only, and that is to carry out the manifestations of Allah's directive as and when He intends it. In other words, Time is responsible for events occurring, when they occurred, as they occur, and when they will occur. By implication, this relates to every event, minute or monolithic, domestic or global, animate or inanimate, human or inhuman.

By definition, History is a branch of study which deals with the documentation of influential events and occurrences in a past Timeline. As such, not every event is regarded as influential, save for occurrences which are either recognized as decisions which affected an outcome, or occurrences which contributed to decisions which influenced an impactful outcome. The study relies heavily on the observer and his perspective of what he considers to be factual and impactful. In short, History is what the Historian writes, and as Tolstoy once said, 'History would be wonderful if only it were true.'

An event is described as an occurrence, a happening, with or without human influence. Holistically, events occur mostly with societal influences, each Macro-event being a result of Micro-events, which are an occurrence of individual actions. Individually, events are isolated occurrences preempted by individual actions, resulting in individual outcomes. Combined, both actions and outcomes conform to individual Macro-events.

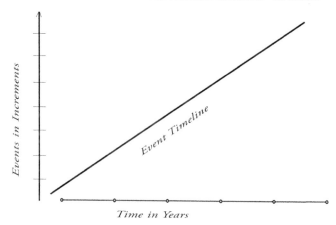

which influenced an impactful outcome. The study relies heavily on the observer and his perspective of what he considers to be factual and impactful. In short, History is what the Historian writes, and as Tolstoy once said, 'History would be wonderful if only it were true.'

An event is described as an occurrence, a happening, with or without human influence. Holistically, events occur mostly with societal influences, each Macro-event being a result of Micro-events, which are an occurrence of individual actions. Individually, events are isolated occurrences preempted by individual actions, resulting in individual outcomes. Combined, both actions and outcomes conform to individual Macro-events.

An example of individual events can begin with the subject waking up and going through a daily routine. If the Macro-event is considered 'going to work' then every Micro-event would include brushing your teeth, washing, dressing, breakfast, and commuting. Each of these Micro-events can further be broken down into a set of Nano-events, every motion and action resulting in a particular outcome. The decisive ability of the subject can impact a Macro-event and its outcome even by a marginal alteration in a Micro, or Nano-event, for example, if one brushes their teeth for a few minutes longer, the resultant outcome may see the Macro-event of 'going to work' resulting in a 'late arrival' with a series of ripple effects further along the Timeline.

Similarly, holistic societal events also occur as a series of individual events, all adding to a particular outcome. For instance, if the holistic event is to 'elect a new president', every individual action contributes to a holistic outcome, from the campaigning politicians, to the observers and possible voters. A fluctuation in individual actions, can result in an alteration of the final outcome, or an individual occurrence which could result in a marginal deviation affecting the decisive vote. If within a particular campaign rally, one individual was to inflict an act of violence, or even make a particular remark, the resultant can be an avalanche of protests and disorder, which can influence the decisive vote of that particular campaigner, thus resulting in a compelling deviation in the overall outcome of the nation's stability or progress.

Regardless of what decisive action influences whatever particular outcome, *all* events, Macro and

Micro must occur with respect to appointed Time, and every appointment is governed by Allah Almighty. This fact is difficult to comprehend by secular minds because not every event is witnessed or documented, at least not from a Historical perspective. The paradox invented from an inability to perceive Time as a governing factor follows in a school of thought similar to that of the tree falling in the forest.

If a tree falls in the forest and no one is there to see or hear it fall, did it really fall?

Similarly, events occurring around the globe, in the present, and in history, are difficult to comprehend, especially when there is no documentation to prove their occurrences.

It is for this very reason that every event mentioned in the Holy Qur'an and Hadith should not be taken for granted. Truthfully, the Holy Scriptures should not be taken for granted at all, but consider that out of the entirety of human history, very specific incidents have been revealed in the Holy Qur'an, and it is without any doubt that we consider these events to be true, because the Holy Qur'an as we know it, is absolute.

As mentioned above, even these events occurred with reference to Time, they occurred by a Passage of Time, and the given fact that they have been mentioned implies that each one of them has a role to contribute in our modern lives.

If we analyze every event with respect to Time, we find that as they occurred, there existed intervals within their concise compositions, intervals of travel perhaps, intervals of tasks or durations of tasks, enactments which by themselves would have taken their own respective Times to manifest, all compounding into the complete tale as it is told.

If we take, for example, the event of Dhul-Qarnayn in Surah Kahf, we find a lengthening of Time between each one of his journeys, and by expounding each destination we also find an occurrence of several minor events, all attributing to a period of Time in Dhul-Qarnayn's life. Refer to *The Three Questions* for an in-depth analysis of the whole event.

As a hypothetical experiment, were we to apply the same event into the modern age, that is, were Dhul-Qarnayn to have been alive in the modern age, we would encounter a strange dilation of Time. His three journeys would be shorter (in Time) given the means of modern transportation today. The minor events would shorten given the ease of communication. Even the task of building the barrier would be eased given the advancement of industry and technology.

The composition of this event, same event— different era, would have shortened in Dhul-Qarnayn's life, thereby giving him the ability to add more such monolithic events in his periodic life.

By this hypothesis, the similitude of other events in ancient times, as compared to modern times, vary in length, giving room for more events to occur, thus creating the illusion of Time moving faster.

The implication of this phenomenon in the modern age is that we can quantify, with near accuracy, the events occurring on a given Timeline are following each other with such rapidity, that Time is seemingly moving faster and faster. An increment of activity in our lives is creating this illusion and delusion, that we begin our day with expectation and it ends sooner than anticipated, every day moving into a month and every month into a year.

As of writing this book, twenty years ago, it would take at least a week to send a letter, and a week later to receive its response, and by the time the entirety of the correspondence had run its course, several weeks would have gone by. In contrast, today, the same conversation can be completed within an hour, two at the most.

Similarly, it would have taken several years in the construction of buildings and houses, whereby today the same composition of construction can be completed in a matter of months. A medical procedure would have endured several hours and a team of specialists, where today, two are sufficient and the procedure is complete within the hour.

To mention but a few, the pattern replicates itself

in almost every situation and every scale with regards to the daily motion of life, yet the actuality of Time remains unchanged. The earth's orbit has not shifted dramatically, nor has the moon's. These Divinely appointed personalities responsible for governing the passage of biological and physical Time, as per the Holy Qur'an, continue to run their course.

It follows, therefore, that the motion of Time, or Time Dilation, is not given by the actual *count* of Time, but by the rapidity with which we, humans, are *spending* Time. The same analogy can be made with reference to currency. One hundred coins several years back would have acquired a magnitude of items, today the same one hundred coins are not enough. Same items, same amount of money— its value dilated.

Similarly, the value of Time has dilated, not its actuality.

With this philosophy, we can better understand the following prophecy, and the impact of the modern age on our lives with relation to Time.

In a Hadith At-Tirmidhi, the Holy Prophet (peace and blessings be upon him) said; *'The hour shall not be established until Time is constricted, and the year is like a month, and a month is like the week, and the week is like the day, and the day is like the hour, and the hour is like the flare of the fire.'*

What the Hadith is describing, and Allah Almighty and His messenger truly know best, is not the rapidity with which *Time will move*, but the *exponentiality* with which *events will occur* with respect to the period or the passage of Time. In essence, condensed within this Hadith, is every single component of the modern age and their impact upon humanity, individually and as a whole.

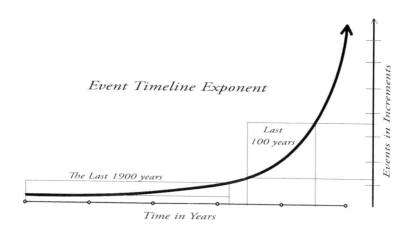

Event Timeline Exponent

Events in Increments

Last 100 years

The Last 1900 years

Time in Years

Take the following into consideration.

A political leader serving eight years, does his share of good and bad. Societal perception is such that even though we acknowledge the 'good' done, what remains dominant to every discussion is the 'bad'. As the eight-year term progresses, one could then surmise a number of inappropriate or unaccepted acts performed as documented events occurring at perceivable intervals of Time.

This means that regardless of how 'bad' the acts are, each one is witnessed, comprehended, analyzed and eventually forgiven by a passage of Time. This is governed by man's intellectual capacity to understand, or 'digest' what just occurred, which is a process that requires Time, and the fallout is then dealt with in accordance, eventually allowing the occurrence to pass with a gradual fade, instead of a rapid and detrimental impact. In short, the wound is given Time to heal.

Following this eight-year term, another leader takes his place and begins to enact the exact sum of events, but in a much shorter bracket of time, such as the first two years, regardless of them being a lesser or greater essence of 'bad'. Nevertheless, the observer objects and rejects with a greater force, because the events occurring have not been fully realized that another adds to the burden, and mainly because Time has not been given a chance to heal.

The similitude is that of a fighter in a boxing ring. If the blows are delivered with longer intervals, the opponent may yet have the chance to recover, reducing the impact and lengthening the fight. If the blows are delivered in rapidity, the opponent has less of a chance to recover, hence the incremental impact, and the quicker conclusion of the fight.

Within this, and the above example, if we consider each act, or blow, as an impacting event occurring on a

Timeline, what ensues is not a shortening or lengthening of Time, rather an exponential increment of events exhausting an existing Timeline, leaving immense room for more events to occur, eventually exhausting all the events destined to occur at a quicker pace.

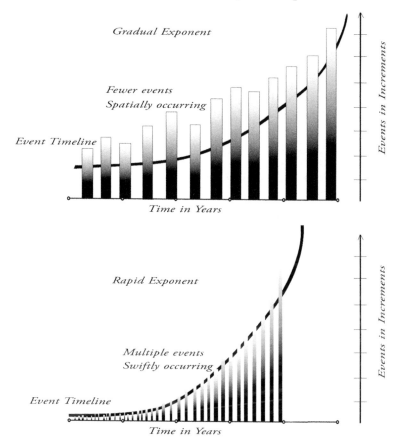

Similarly, the reality of life on earth for the entirety of mankind has but a finite measure of predestined events to occur, every event like a domino in a chain, leading to the last. A fair measure of Time between each piece toppling, lengthens the existence of a Timeline

85

and prolongs the advent of the end.

However, in contrast to Religion emphasizing patience, man has proven himself to be a hasty creature. Ever struggling to invent new ways to acquire new things with haste. Traveling swifter, acquiring faster, communicating quicker. It does not take much to visualize the exponentiality with which we have hastened the manifestation of the modern age, hence ushering in our finalities.

Take, for instance, the global human population (as illustrated). Every event is as a direct result of individual inputs, all collectively contributing to a 'Global Stream' of events. Chiefly, these are governed by trends, such as technological trends, lifestyle trends, political trends, and so on. The incremental activity of individual actors (humans) is heavily dependent on the Motion of Time, and with every increment there is a direct and realistic impact, influence, and demand on every other 'human' essential as per the laws of human Cause and Effect. Food, Industry, Governance, Academics, and so on, are all driven by human influence. This also includes the ripple and consequential effects of every event, such as waste, pollution, disputes, and wars.

The 'everything' as a direct and indirect resultant of human demand continually contributing to the 'Global Stream', raises the exponent of the eventuality, as well as drastically and dramatically contributing to the illusion of Time moving swifter. The graph below depicts a reality of this phenomenon.

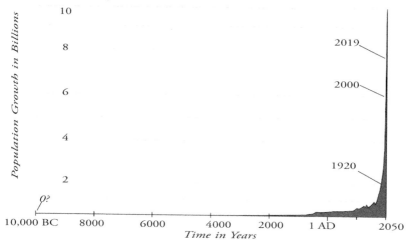

Now, the secular world would 'scientifically' claim that this exponential growth is as a result of human intellectual advancement through evolution, and therefore inevitable given the rapidity with which human intelligence is growing. Unfortunately, this presents a globally catastrophic effect, in that the finite sustainability of the earth is also being exhausted exponentially. With statements such as 'global warming', 'sustainability', 'carbon emissions', 'energy consumptions', and what-have-you-not.

The truth of the matter is that every demand for sustenance is made on a material scale. It is either a demand to sustain technological needs, comfort needs, or consumption needs. Rather, should we say, *unnecessary* technological needs, *lustful* luxurious needs, and *gluttonous* consumption needs.

These effects should not be ignored, but unfortunately, nor can they be reversed. The finger can be pointed at every one of the terminologies above, but for the large part, it can only be pointed towards ourselves. They are the irreversible effects of our own inhibition of dissatisfaction and greed.

Exponentially, this is depicted as such;

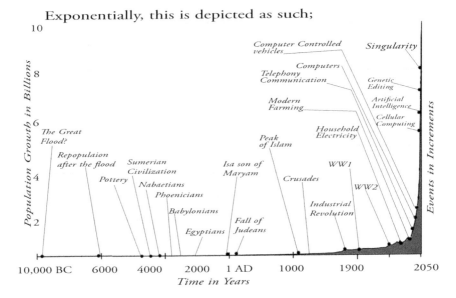

As far as the earth is concerned, it is well within its capacity, as per Allah Almighty's Decree to sustain *all* life on the planet, including animal and plant. Note that animal and plant populations far supersede the human population by *nth*-factors, yet we are the only species with an insatiable hunger and thirst. Unfortunately, rather than recognize its limits, mankind is ever pushing forward and pursuing every possible means to increase its capacity to consume. Here we see the exponential growth of Artificial Intelligence and Robotics. Historically, man used animals to increase productivity, and now we are using machines that can move faster, stronger, and hence produce more for man's consumption. This results in another exponentiality, with man's inability to 'keep up'

with technological advancement, and so results the need to infuse man with machine, hence loosing its humanity.

As of our current, and utterly irreversible position in Time and Space, the balance is immensely biased on the side of negativity. Hence the reason for *Aakhira* arriving in the manner as it currently is. With earthquakes, with tyrants, with godless societies, empty ideologies, wars, politics, famines, droughts, disease, the list can go on exponentially.

Taking all the above into account, can we thus explain our world from an Islamic viewpoint?

One particular advent of the end, is the Dajjal. A widely misidentified character, chiefly because of his talent in espionage and deception.

The Holy Prophet (peace and blessings be upon him) said, *'The Dajjal will live on earth for Forty Days. A Day like a Year. A Day like Month. A Day like a Week. And the rest of his days will be like your days.'* (Bukhari, Muslim)

Plotted on a graph, his Timeline would look like this;

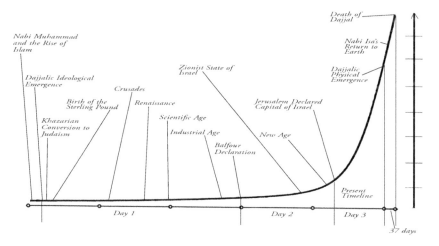

With all the above in mind, is it at all complicated to realize that he who has been ordained to arrive and distort our existence through colossal trials and tribulations, has already arrived, albeit ideologically?

The greatest threat is not that the trials are overburdening. It is that we are being swept off our feet, such that before we even realize it, we have symphonically fallen into the trap of the Dajjal, thus dealing with the outcomes on a greater scale with a greater cost, adding to the burdens of Trials and Tribulations.

All this occurring simply because we are so obsessed with our material lives, we cannot take the time to understand the actuality, sublimity, and Divinity of Time to enable us with the Knowledge, Guidance, and Patience to deal with the modern age on a spiritual level.

TIME IN DIMENSION

Thus far we have understood the link between occurrences and Time, and now we can look at the overlaying nature of Time over occurrences.

This overlaying nature is given by the manner in which a measure of Time is direly necessary for an event to occur, for instance, if an event takes a complete hour to occur, it can only take that particular hour, no less. In this instance, until the entirety of that event has occurred, can the next event take place, and so on.

Additionally, some events can occur as dependents of each other with equivalent Passages of Time, and some events, however alike, can also occur independent of each other, neither one conflicting the other's Time.

This overlaying nature of Time governing specific

events in their specific nature of occurrences is regarded as the Dimensionality of Time.

In order to understand this concept, one must first withdraw from the conventional perception of 'Dimensionality', not as coordinated structures with quantifiable measurements, but as *'States of Existence'*. This is a crucial perspective in the study of Time, because as we have so far understood it, Time is not a quantifiable object.

Einstein's theory of Special Relativity further sheds some perceptive on the concept that Time is also multidimensional. It can transcend through the ages, it can transcend through the heavens, each heaven a dimension by itself. Through Einstein, science has come to understand, over the past century, that space and time are not separate entities, but one and the same— Space-Time.

Within material study, science has thus far identified three observable dimensions in three-dimensional space, the physical world we occupy, the expanse of Outer Space, and the world below the atom, that is, the Quantum Realm. Beyond three-dimensional space, physics becomes tricky because human perception lacks beyond three dimensions. Further to either horizon, for instance inside a Black Hole in space, or in the Quantum Realm, the Laws of Physics transcend into chaos, because Gravity, which governs the principle of our perception (coordinating up, down, left, and right) loses its integral value of Speed, Displacement, and Vector.

However, we cannot deny the existence of Quantum particles, nor the intensity of Black Holes, and in order to give physics the opportunity to make sense, the notion of a 'Multi-verse' was introduced. The notion has been hypothesized and debated over numerous branches such as 'alternate universes', 'quantum universes', 'inter-penetrating dimensions', 'parallel dimensions', 'parallel worlds', 'parallel realities', 'quantum realities', 'alternate realities', 'alternate dimensions' and 'dimensional planes'. Of course, neither of these hypotheses can be proven right without substantial evidence, but within the context of the hypotheses, one governing principle remains resolute through every instance, regardless of the validity of physics within that instance. Time.

The focus of this study is not to explore nor validate either of the hypotheses above, but to understand the essence of Time within dimensional perspectives. If Time is relative, it bears authority to every occurrence in Quantum Realms and Celestial Realms within varied States of Existences, or possible existences. Therefore, if Time is relative within a closed system, such as on earth, the solar system or observable universe, it should also adhere to relativity within alternate dimensions with alternate Timelines, or alternate outcomes.

Let us first understand dimensionality as is explained

by the Holy Qur'an.

Muslims, and by extension even non-Muslims, cannot refute the possibility of otherworldly beings in other worlds. Angelic or Demonic. *Malaikah* or *Jinn*. The issue is not in rejecting otherworldly presence, but in understanding their actuality in Time and Space. It becomes complex when we try to confine them into coordinates, whereas the simplest response is that of dimensional existence.

As explained;

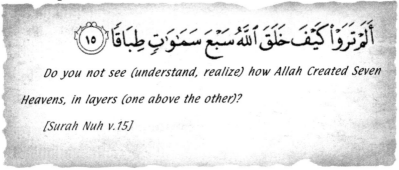

Do you not see (understand, realize) how Allah Created Seven Heavens, in layers (one above the other)?

[Surah Nuh v.15]

In its simplest definition, the word *Tibaaqan* is derivative of *Tabaqa,* which literally translates to 'layers', but also translates to 'stratum' which denotes 'parallel multi-layering', and in physics, it denotes 'dimensionality'. The existence of multi-layered worlds is evident from the verse, but so is 'that which separates the layers'. For example, between the physicality of the earth and space, there is an atmosphere that separates the two. With respect to Time existing within all these layers, however, we cannot relate to a *physical* separation, as such, we can only identify the 'separating' or 'dividing' element as a 'dimensional barrier between dimensions'. In the eloquent language of the

Qur'an, 'dimensionality' is often described allegorically with the word *'Hijab'* (veil), such as;

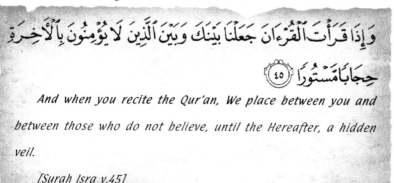

And when you recite the Qur'an, We place between you and between those who do not believe, until the Hereafter, a hidden veil.

[Surah Isra v.45]

The composition of this 'veil' matters little to us, so much as its existence as a partition between two worlds. It is from this verse, and a multitude of several others, that we can understand how a partition between the Angelic, the Jinn, and Human realms exists.

As per the description of the verse above, a veil between a believer and non-believer does not split the world into two physical locales. The physical planes of both worlds continue to exist as they exist. The corporeality of both worlds endures on a continuum independent of each other. The world of the non-believer remains materialistic, while the world of the believer reveals a more spiritual marvel to the human self.

Irrespective of either worldly composition, Time continues to impose its will. In this context, the world of the non-believer only reveals a mere Mechanical Count of Time, whereas the world of the believer reveals a more

Sublime and Divine Understanding of the Passage of Time.

In similitude, the realms of the Angels and Jinn, in contrast to the realm of man, have their own dimensionality of Time passing, independent of each other, all partitioned by a 'veil'.

It is important to note that regardless of either of the mentioned dimensions, neither has left the *physical* confines of the earth. Several hypotheses have arisen over the past, propagating Parallel Worlds of Parallel Universes with Parallel Earths, all bent on validating themselves as 'true' in order to rationalize the physics. These hypotheses attempt to define an alternate set of elements (such as an alternate set of human beings, exact duplicates of ourselves), all abiding by a multitude of alternate occurrences, with alternate decisions and hence, alternate outcomes. Science would disagree, but the concept of 'Parallel Universes' with a parallel set of possibilities and existences is absolutely fictitious, namely because the Holy Qur'an does not endorse any of those hypotheses.

The Holy Qur'an continues to impose the statement 'Heavens and the Earth' in no less than one-hundred and eighty times, and nowhere does it speak of 'parallel worlds' in *any* form. Only Allah Almighty Knows best His Creation, so what need is there for man to speculate over fictitious belief for the sole purpose of 'making things fit'?

Consider the following verse;

يَٰمَعْشَرَ ٱلْجِنِّ وَٱلْإِنسِ إِنِ ٱسْتَطَعْتُمْ أَن تَنفُذُوا۟ مِنْ أَقْطَارِ ٱلسَّمَٰوَٰتِ وَٱلْأَرْضِ فَٱنفُذُوا۟ لَا تَنفُذُونَ إِلَّا بِسُلْطَٰنٍ ٣٣

O ye of the Jinn and Men, if you (think) you have the ability to venture beyond the frontiers of the Heavens and (penetrate the depths) of the earth, then proceed; Ye will never venture forth, save without Our authority.
[Surah Rahman v.33]

Allah Almighty is addressing human and Jinn collectively, implying a very simple fact; both creatures are upon the same platform— earth. Within this verse are contained the three observable dimensions of three-dimensional existence, namely the neutral coordinates of the earth (*denoted by addressing Man and Jinn*), the edge of the observable universe from the perspective of the earth (*the frontiers of the Heavens*), and the Quantum realm (*the depths of the earth*).

It is widely understood that *we* cannot perceive the Angels in their Angelic Dimension, nor can we perceive the Jinn from our designated *Human dimension* on the earth, as explained;

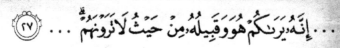

...Verily, he (Iblees) and his tribe (the Jinn) can see you from where you cannot see them... [Surah Al-A'araf v.27]

The implication is that aside from the Seven Heavenly dimensions, the Seven *Samawat*, there exist also inter-dimensions within each *Samawat*, thereby implying that there also exist inter-dimensions within the dimension of the observable universe and the earth.

Within our observable universe, we can explain the mathematics of dimensions using the literal definition of 'dimensions' as plottable coordinates on various axes. In three-dimensional space, we use the *(x)*, *(y)*, and *(z)* axes to plot the coordinates of an object in three-dimensions, because the laws of physics are still the governing principles, and by extension so are the laws of perception. From the observer's viewpoint, a three-dimensional object can be identified to be x-coordinates across the observable plane, y-coordinates above the plane, and z-coordinates away from the observer.

A one-dimensional object cannot perceive a two-dimensional object, but the reverse is true. A one-dimensional object only has one axis, either *(x)* or *(y)*. It exists only just, as a point in Space-Time, able to perceive or move along only *one* of its designated axis.

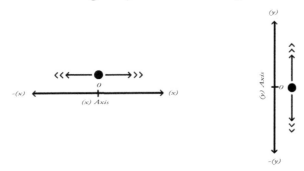

A two-dimensional object has both an *(x)* and a *(y)* axis. In this case, the two-dimensional object can acknowledge the one-dimensional object, by altering its perspective while able to move along both axes, *(x)* and *(y)*.

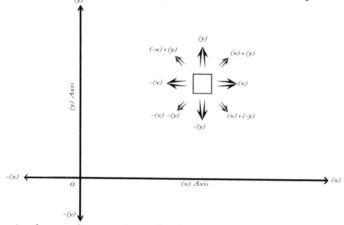

A three-dimensional object, such as ourselves, have three axes *(x)*, *(y)* and *(z)*. We can perceive both the first and second dimensions by altering our perspective while being able to alter our vectors along all three axes.

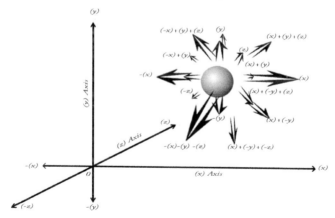

Beyond the third dimension, our perception is very like the first two objects (in the first and second dimensions) that cannot perceive the third dimension. Similarly, we cannot perceive a dimension greater than three. In actuality, aside from perception, we cannot biologically exist in two, or one-dimensional space either, because we are three-dimensional objects. In order to do so, we would have to take the shape and form of a lower dimensional object, a feat impossible because humans are defined life-forms. Additionally, we would also have to conform to the concept of Time and Space in these alternate dimensions, both of which are beyond our three-dimensional capabilities.

Similarly, in the realms (dimensions) of the Jinn, as well as the Angels, Time and Space manipulate and interweave in a manner that is incomprehensible to us. We cannot perceive either of these beings in shape, form, or composition, until they cross over and enter *our* dimension in the form of something that adheres to three-dimensional space. Typically, the form of human or animal.

The entirety of the above study is to bring to perceptive the existence of Spacial Dimension, and in accordance with Space, so too does Time have a dimensionality and inter-dimensionality. With regards to realms other than human, our perception of Time bears no significance, but one must not ignore the possibility of dimensionality even within our human realms. What this means is that the dimensionality of Time, as an intangible element, is not limited to the dimensionality of Space, as a tangible element, and the Fluidity and Relativity of Time enable it to have an inter-dimensionality of its own. We can study this particular concept with relevance to ourselves.

In this case, we have to use allegories to enable an understanding of this concept, and an excellent analogy can be made by using archery bows.

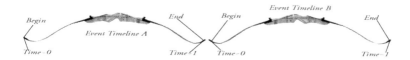

If we place the bows side by side, as illustrated, one tip extending into another, the bows become *Linear* (Sequential Time).

One portion begins and ends, followed by another portion. This can be understood as the Unidirectional Passage of Time (always moving forward) on earth such as when describing the corporeality of human existence on earth.

For example, a child cannot be an adult, nor an adult a child, nor both at the same time. Infancy must occur first, followed by childhood, then adulthood, sequentially. This concept of Time as always moving forward in sequential steps can be acknowledged in nearly every aspect of earthly motion.

In another example, we cannot occupy two instances of space, hence two occurring events, in the same instance of Time. We must complete what needs to be completed in one space before moving to the other. Hence, the Passage

101

of Time flowing in this space from beginning to end, and another Passage following in the next space.

What this means is that, with events occurring to a respective Timeline, some events have to take place in their entirety before the next can have its opportunity. Event (A) then (B) then (C) and so on. Neither event can 'overlap' the other, or 'overtake' the other. This Sequential Passage is such that human existence is ever confined to a forward motion, with whatever done is done, and whatever has to happen will happen. Even though an element of 'multitasking' is within human capabilities, to a very refined count of Time, individually, every task is sequential in every instance.

This motion of Time is the immediate and most evident perspective of Time in every individual's life. The underlining principle is that the actual sequence of events does not matter, (A) then (B), or (B) then (A), but that the motion of Time through either sequence, or set of sequences is ever in a Linear, Sequential, and a forward direction.

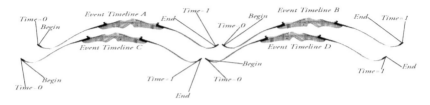

If we place the bows one on top of the other, as illustrated, we form a *Layer* (Parallel Time).

This is a simple concept of duplicating or multiplicating

102

Sequential Time, given by two or more events, similar in nature, all occurring in the same instance with equivalent proportions. Typically, this motion of Time is equivalent in the same instance, but separate in Space, still adhering to a forward motion such as that of Sequential Time.

An example of this duplicity, or multiplicity, can be given by a marching procession, whereby each individual's march forward is equal to the next, hence each one's Timeline being Parallel to the other.

Another example can be given by workers in an office, or an industry, whereby their working schedules would begin and end within the same intervals.

A more profound example can be seen in nature, such as that of similar flowers blooming within the same intervals, or weather occurrences such as rain, or snow, occurring in the same instance but in different spaces, or different parts of the world.

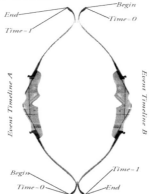

If we place the bows in adjoining arches, both tips touching together, as illustrated, to form a 'circular' shape, it becomes a *Loop* (Cyclical Time).

103

This can be used to describe the Passage of Time on earth such as when describing a Cyclical occurrence, like the Four Seasons. Such kind of Time can also be applied to reoccurring earthly motions, such as Seed Germination, or the cycles of sleep and wakefulness.

This motion of Time is akin to its Linear counterpart in the sense that although most cycles are repetitive, they are also always in a forward motion and sequential in their own respect. In the example of the seasons, the sequence of events is given by Spring, Summer, Fall, and Winter, all sequentially following each other, but also recurring when the cycle is complete.

With human activity, the motion can be given to any series of events directly or indirectly influenced by human interaction.

An example of such can be habitual activity or daily routine. Muslims who pray five times a day, go through such cycles in every prayer, albeit limited to Two Cycles in *Fajr*, Three Cycles in *Maghrib*, and Four Cycles in *Dhuhr*, *'Asr* and *'Ishaa*. Similarly, when performing Hajj, *Tawwaff* around the Ka'abah is a cyclical motion of Time, albeit limited to Seven Cycles.

Holistically, the Muslim Ummah goes through daily cycles of *Salaat*, and yearly cycles of *Zakaat*, *Hajj*, and Fasting in *Ramadhan*. Within the month of *Ramadhan*, every fast is a repetitive cycle in and of itself, with a ritualistic motion through sunrise, sunset, and overnight.

104

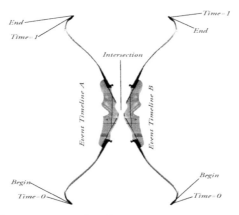

Finally, if we invert the bows so that their tips point *away* from each other, as illustrated, and only the curves touch together at their apexes, we form an *Intersection* (Dimensional Time).

This is where two or more independent occurrences, or Timelines, run alongside each other, neither influencing the other except where an intersection takes place. Such motions of Time are much more common as they can incorporate both Linear and Cyclical times in a multitude of levels, thus adding to the complexity of Time. Every object on a holistic Timeline, man, animal, plant, or Jinn, has an individual Timeline ordained by Allah Almighty, each one taking its own course.

As daily life endures, interactions between all the above elements, including inanimate objects, create intersections in each Timeline. It should be noted that these intersections can also occur between Spacial Dimensions, such as between the Heavens and the observable universe, between the dimensions of Man and

105

Jinn, and this would then explain the strange manner in how the current dimension of the Dajjal is interacting with our biological dimension by way of which his influence is already upon us.

Both theoretically and practically, a breach between dimensions requires a two-way interaction, one which is the sender of information, the other which is the receiver. Without delving too deep into the matter, the simplest explanation is that *he,* in his dimension, is the imposer of his ideology and influence, and there are those in our dimension who are the implementers of that ideology, with every actor, action, and interaction, a Timeline by itself. The actuality of the matter is multitudes more complex when we look at the formation of Time and Events in layers upon layers, sequences after sequences, and continuous intersections, all relative to one another.

TIME
IN THE
GODLESS WORLD

As we continually, almost unknowingly, become caught up in the complexities of life, we often tend to forget that we are ultimately dependent on Time, and Time possesses a reality that it is independent of mankind. As is said, Time waits for no Man. The clock tells us what Time it is, but never tells us what Time *is*.

The Holy Qur'an tells us that Allah Almighty has created the Sun, Moon, and Stars with designated duties, among which one of them is to enable us a measure of Time.

The question then posed is this;

Along with the components of measurement (Sun, Moon and Stars, orbit of the Earth), is the count of

Time also ordained by the Creator, or is this something we have to deduce for ourselves based on our own perspective?

The sun determines the seasons and years, the moon determines the months, the earth's rotation determines day and night, and the hours of the day and night are all determined by the harmonious waltz of the earth with the moon and the sun. So again we ask, is the measure of Time an act of Divinity, or an instrumentation of man?

The Holy Qur'an says;

إِنَّ عِدَّةَ ٱلشُّهُورِ عِندَ ٱللَّهِ ٱثۡنَا عَشَرَ شَهۡرًا فِى كِتَٰبِ ٱللَّهِ يَوۡمَ خَلَقَ ٱلسَّمَٰوَٰتِ وَٱلۡأَرۡضَ مِنۡهَآ أَرۡبَعَةٌ حُرُمٌ ذَٰلِكَ ٱلدِّينُ ٱلۡقَيِّمُ فَلَا تَظۡلِمُوا۟ فِيهِنَّ أَنفُسَكُمۡ وَقَٰتِلُوا۟ ٱلۡمُشۡرِكِينَ كَآفَّةً كَمَا يُقَٰتِلُونَكُمۡ كَآفَّةً وَٱعۡلَمُوٓا۟ أَنَّ ٱللَّهَ مَعَ ٱلۡمُتَّقِينَ ﴿٣٦﴾

Indeed the number of months with Allah is twelve months (inscribed) in the register of Allah (in his Ordinance) when He created the Heavens and the Earth, of them four (months) are sacred; That is the Religion (ordained) rightful (and True); So do not wrong yourselves therein...

[Surah Tawbah v.36]

As an interpretation of the above verse, the Holy Prophet (peace and blessings be upon him) said, 'The division of Time has returned to its original form which was current when Allah created the Heavens and the

Earth. The year is of twelve months, out of which four months are inviolable; Three are in succession *Dhul-Qa'dah, Dhul-Hijjah,* and *Muharram,* and (the fourth is) *Rajab* of (the tribe of) Mudar which comes between *Jumada Ath-Thaniyah* and *Sha`ban.'* (Bukhari)

It should be noted that the Hadith was documented as part of Nabi Muhammad's final sermon, and the verse was revealed to undo the ideological destruction of the pagans, by which the cyclical periods of *Hajj* and *Umrah* had become distorted. The *Hijri* calender Divinely Decreed was thus restored, with the four scared months prohibiting any frontal Jihaad against the enemies of Islam during these months.

Within the context of this study, from the verse and the Hadith, we can understand that He who created the Sun, Moon, and Earth, and ordained their passage as a measure of Time, among other appointments, He who Created Time itself, also ordained a governance and an appropriated Passage of Time to abide by what the verse above declares. The Solar Calendar which we so blindly follow in the modern age is not sanctioned nor advocated by Islam.

One must take a moment to try and understand why Allah Almighty made the outright declaration to abide by the Lunar Calendar of Hijra. What is the significance of taking the last fourteen hundred years in its appropriate format, instead of abiding by the widely used Gregorian Solar Calendar?

How have we become so lame, so ignorantly accepting and adapting this modern age, without a moment's thought of what Timeline we are following?

The Impostor has used every possible trick to destroy the harmonious bond between Time and Life, as was and is ordained by Allah Almighty. The modern world has sought to corrupt every perception of Time, resulting in a crippling blow to our capacity to think and measure Time in any other way except for the mechanically instated way.

Throughout history, variances of calendars have been generated to keep a count of Time, with every civilization subsequently deriving their own methods of count, some based on the Solar Cycle, some on the Lunar Cycle, and some even on Zodiacs.

One such example is of the ancient Mayans, to whom the numbers 13 and 20 were sacred (in their own belief system). They had a series of calendars, beginning with the *Tzol'kin* calender which was a measure of social and spiritual events, and in Gregorian counts, it had 260 days to a year (13 months with 20-day periods). For agriculture and seasonal counts, they had a calendar called *Haab,* which had 360 days plus five unnamed days called *Wayeb* (nameless days) which they considered 'unlucky' or 'deprived of Godly presence'. Both calendars lined up with each other every 52 years, regarded as one cycle. They extended this cycle as such— A *Kin* is a day. 20 days from the *Tzol'kin* was *Uinal* (month). One year from the *Tzol'kin* was a *Tun* of 360 days. 20 years was a *Katun* of 7200 days. 20 *Katun* was a *Baktun* of 144,000 days.

Why is this relevant?

Because the entire cycle of 13 *Baktun* (5125 years) coincided with the date of December 21st, 2012 on the Gregorian calendar, a date that was widely propagated to be the 'end of the world'.

Again, why is this relevant?

Because of futile attempts at trying to calculate and quantify Time, in order to uncover *'I'lm ul-Ghaib,* knowledge of the unknown and unseen, and by extension, knowledge of the future. Man's most dire and curious need has always been to determine what would come to pass, because man's most restless state of existence is one without knowledge.

There is a very profound reason why Nabi Muhammad (peace and blessings be upon him) gave us prophecies of the end of days with such intricate detail and in a manner that included a crucial aspect of Time, and it was not to enable us to calculate how long we have left, but to actualize a reality of things by penetrating the concept of Time which would be otherwise be artificialized by the Dajjal.

The widest misconception of the Dajjal is that he is supernatural being with mystical supernatural powers, when in truth he is just as human as the next man, with one unique ability, and that is to disguise himself as an Impostor. He disguises himself not by *wearing* a disguise, but by way of the oldest trick in the book known as a 'slight of hand'. He creates a system which deludes the masses, clouding their judgment and crippling their ability to think and ponder, hence unable to decipher what the truth really is. One of the most fundamental pillars he has destroyed has been our perception of Time ordained by Allah Almighty, preventing us from taking a moment to *think,* thus leaving us defenseless and unable to see the signs leading to our destruction.

The Holy Qur'an says;

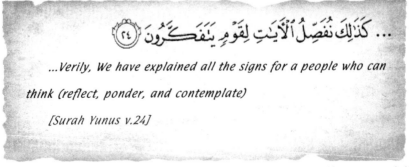

...كَذَٰلِكَ نُفَصِّلُ ٱلْآيَٰتِ لِقَوْمٍ يَتَفَكَّرُونَ ﴿٢٤﴾

...Verily, We have explained all the signs for a people who can think (reflect, ponder, and contemplate)

[Surah Yunus v.24]

For what has the modern, secular, godless world done of Time?

The 365-day year, 12, 30-day months, famously called the Gregorian Calendar introduced by Pope Gregory XIII in 1582 AD, which was proposed to refine the centuries old Julian Calendar (introduced by Julius

112

Caesar from January 1st 45 BC) originally adapted from the ancient Egyptian calendar of exactly the same structure. The twelve months of the year have been named thus, just as the seven days of the week have been named, to replicate the names of pagan deities. Such deities extend from the far reaches of ancient Egypt to the steppes of ancient Rome and the columns of ancient Greece.

Deliberated thus upon the entirety of mankind, so that with every mention of the date, we invoke nothing short of a pagan deity.

January is named after Janus, the god of beginnings.
March is named after Mars, the god of war.
April is named after the Greek goddess, Aphrodite.
May is named after the Greek goddess, Maia.
June is named after the Roman goddess, Juno.
July is named after Julius Caesar himself.
August is named after his son, Augustus Caesar.
Saturday is named after the Roman god, Saturn. (Saturn's-Day)
Sunday is named after the Norse sun-god, Sunnu or Sol.(Sol's-Day)
Monday is named after the Norse moon-goddess, Mani.(Mani's-Day)
Tuesday is named after the Norse god of war, Tyr. (Tyr's-Day)
Wednesday is named after the Norse god, Odin. (Odin's-Day)
Thursday is named after the Norse god, Thor, son of Odin.
(Thor's-Day)
Friday is named after the Norse goddess Freya, wife of Odin.
(Freya's-Day)

Similarly, all the major the celebrations of Christmas, Valentines, Easter, and Halloween, all derive pagan traditions either from pagan Rome, Egypt, Celtic or Nordic traditions, having little or absolutely nothing whatsoever to do with Christianity in its True Faith and Form. This was no mere accident, yet it has so craftily escaped the wits of many who claim to be scholarly spiritual in Christian and Judaic Religions.

The architecture is such that even though we may be following our respective religions, all of us under Abrahamic Faith eventually adhere to some form of paganistic rite, often unknowingly and unwarily. In doing so, we tread ever so deeper into a polytheistic and mechanical way of life, every step leading to a mechanical evolution of mankind (*The Abyss of the New World*).

To the modern man, a day hardly ever ends with the beauty of a sunset, nor does it begin with the marvel of sunrise. Instead, a day ends and begins at midnight in a completely irrelevant, inconsequential and utterly meaningless fashion. Of what *use* is the beginning and end of the day at midnight when the populous is fast asleep? *What*, pray tell, is the *purpose* of 'Daylight Savings'? Or even the need to continuously refine a precise count of Time?

A new month no longer begins with the beauty of a new moon and end with the grace of the next, as nature has ordained. Instead, the month is mechanically parted into thirty or thirty-one equal days just as every day is parted into twenty-four equal segments. Whatsoever happened to the grace of dawn, the chill of morning, the sultry of noon, the waning of dusk into fading twilight and eventually the dark of night. We now measure the extent of our lives in mechanical advances, counting every bit, right down to the seconds.

Simply put, a miserable quest in a mundane pursuit for the material over the sacred.

The modern world that many of us so gracefully and ignorantly embrace has been designed for the sole purpose of entrapping us in the 'moment'. So that we think of naught but that next business deal, that big game, that next milestone, that drama, that 'like' and that 'comment'. Images flashing across a screen, both as large as a room and as small as a palm, with a rapidity that distorts, diminishes and decays our capacity to think, ponder, and reflect. They reduce us, and further reduce our children, to mindless drones, day in, day out. Yesterday fades into nothingness without a mark on our conscious, and tomorrow remains but an extension of today's fantasy.

The consequence being that humans, whether Muslim or not, become a mentally incapacitated herd of cattle, worse than cattle (*Al-A'araf v. 179*), unable to comprehend the present, unable to relate to the past,

115

and therefore utterly blind in one eye.

As the Dajjal is blind in one eye.

We fail to understand Time as it has been in History, as the events have unfolded in the past, hence unable to recognize the unfolding of a mysterious endeavor, a vile agenda, in the world today.

Allah Almighty has not given us an outright description of Time for such paramount knowledge would become meaningless to us were we to acquire it without struggle. Striving to acquire knowledge bears a fruit so sweet, its likeness cannot be of anything definable. Allah Almighty has not forbidden us to pursue this knowledge, and therefore whosoever, scholar or otherwise, dismisses this quest has sourly misguided himself and others. Woe unto those who misguide others from truth and righteousness.

To dive into the depths of the Holy Qur'an and arise with magnificent pearls of Knowledge, one must look with both eyes, both external and internal. Scattered throughout the Holy Book are clear locations of these pearls and how to unearth them. To do so, one must respectfully shed the godless structure of a godless world, and embrace a Divine structure of Time as has been Divinely ordained.

TIME
ON THE
DAY OF JUDGMENT

The most intriguing element of any book, any story, is it ending. Its climax. It constitutes the finality of all documented events with a perplexing question on edge— what will happen?

Man's existence could be little else but a very long book, with a phenomenal beginning, a perplexing structure and plot, and an impending climax, unlike anything ever witnessed. So much so, that its foreshadow, in and of itself, is puzzlement beyond comprehension.

The questions then posed, are these;

Can we recognize the signs without the need to comprehend them? What constitutes comprehension? Is it all but reading words, or is there a deeper effort

117

required? If so, what are the guidelines, or ropes, for delving deeper? How deep can we go? Can we eventually decipher the absoluteness of the Last Day, down to the Last Hour?

The Last Hour may as well be the Final Chapter in the Book of Man, but after that... what?

It is important to note that the finality of all things will not occur in one Chapter. The Last Day (*Yaum ul-Aakhira*), the Last Hour (*As-Saa'ah*), and the Day of Judgment (*Yaum ul-Qiyamah*) are all separate entities by themselves. Let us begin by understanding what these different chapters mean.

What is the difference between the Last Day and the Last Hour? Does this day consist of twenty-four hours, equally stretched out to accommodate all the events ordained to take place? Will the Hour consist of sixty equivalent-minutes elongated to accommodate the Final Statements of life on earth and in the universe?

If we attempt to answer these questions from our perception of 'Earthly Time', we are only fooling ourselves with delusions and speculations, when Allah Almighty *has* given us the guidelines to acquiring and understanding this knowledge.

Understanding Time Divine enables us to acknowledge that a Divine Day is not a twenty-four hour period. It could be an 'Age', 'Epoch', 'one thousand years', 'fifty-thousand years', or even hundreds

118

of thousands of years, all variantly incomprehensible to our reckoning (*Al-Hajj v.22, As-Sajdah v.5, Al-Ma'arij v.4,* and even the Bible *Psalms 90:4*).

Given the above, we can now understand that the Last Hour cannot be a sixty-minute period, but a *period that marks the pinnacle of the Last Day.* The Hour when monumental changes occur, the Hour that marks a series of climactic events, regardless of its duration, it is the Hour when a Divine Singularity takes place, from which there is no return, as a Singularity in definition marks a point of no return. Following this finality, everything of Allah's Creation will be made to perish, following which will begin the next age, or the Day of Judgment (*Qiyamah*).

We can use science, not to quantify the Last Day and Last Hour, but to allegorically define them, in order to better comprehend their Divine Nature.

In Astrophysics, a Black Hole is a mass in space with a gravitational pull so dense, that not even light can escape it, which is why it appears to be 'black'. Its composition eludes scientific logic, because the Laws of Physics cease to make sense in a Black Hole, namely the Laws of Gravity, Motion, and the Laws of Thermodynamics. At the center of a Black Hole is a Singularity, a point where all drawn matter converges, and every Singularity has an 'Event Horizon'. This is the outer 'shell' that encompasses the Singularity, which

119

in and of itself, is also a point of no return. Aside from matter, there are two main elements at play with regards to a Black Hole.

One is Light, and its inability to escape the Black Hole. It becomes invisible and immeasurable because of the 'pull' on it. This is a phenomenon without explanation, namely because Light has no mass. As witnessed, the Laws of Physics disintegrate into chaos, and one portion of the chaos is to explain how something without mass can be affected by the immense 'draw' of gravity.

The other is Time, namely because Time in its forward motion is the participating governor of the entire process of the Black Hole, but is also one of the components that is 'drawn' into the Black Hole. Theoretically, if a human was drawn into the Black Hole, past the Event Horizon, not only would his mass and physical structure be elongated, but Time itself would seem to move faster and faster, the closer he moves toward the Singularity. This must not be confused with the theory that *outside* the Black Hole, on the edge of the Event Horizon, Time theoretically 'slows down'.

The Last Day, therefore, can be allegorically described as a Black Hole, with the Event Horizon marking its beginning, and the Last Hour, can be allegorically described as the Singularity, marking the end of the Last Day. What happens beyond the Singularity, in this case, what happens after the Last Hour, cannot be fully quantified, but can be comprehended from the Qur'anic definitions of the Day of Judgment, also named the Day of Resurrection.

Using this analogy, we can hence understand how the Story of Man unravels. With the Black Hole, we can only observe what is happening, or what could happen (anticipation) from the outside, and everything remains but a theory or hypothesis until it is actualized. To fully grasp the eventual climax, and to understand what is really happening and why it all happened, we have to be *inside* the Black Hole. The Event Horizon is also considered a point of no return because Time itself is drawn into the Black Hole, and so to escape it, one would have to travel back in Time, an impossibility in and of itself. It is here that we realize, as the events circumferential to our lives unfold and the Signs become more and more evident, that indeed we have thus entered the Black Hole, or the Last Day, in fact, we have already ventured far beyond the 'Event Horizon' and much closer to the Hour than presumed.

Perhaps it would appear to the reader as not so. It is quite apparent, judging by the state of the world today, that the evidential signs have not yet fully been realized by so many, and it would take more than a few pages to fully explain what is meant by the above statement. We urge the reader and researcher to consult *The Three*

Questions for a more comprehensive response, but to surmise a moderate explanation, we can spare a pair of paragraphs.

Of the numerous foretokens of the Last Hour, the Holy Prophet (peace and blessings be upon him) highlighted Ten Major Signs, namely;

The release of *Ya'juj* and *Ma'juj*, The arrival of the Dajjal, *Ad-Dukhaan* (Smoke), *Daabbat ul-Ardh* (Beast of the Earth), The Rising of the Sun from the West, Three Quakes (one in the east, one in the west, and one in Arabia), A fire out of Yemen would drive people to their place of assembly, and The Return of Nabi Isa son of Maryam. (Bukhari and Muslim)

From our rigorous studies, we have confidently identified that *Ya'juj* and *Ma'juj* were released during Nabi Muhammad's time, as well as commencing the Dajjal's First Day like a Year. We have identified the rising of the sun as the might of the Western Empire. We have identified the Beast of the Earth as the Zionist Force, and as of writing this book, we have yet to identify the remaining signs. We can anticipate *Ad-Dukhaan* as a possible nuclear fallout of the impending *Malhama* (Great War), and the return of Nabi Isa to be in a matter of years as the Great War and the Dajjal's Third Day like Week (our current position in Time) complete their course.

We urge the reader to consult *The Three Questions*

to fully understand how we have arrived at these deductions, and Allah Almighty truly knows best whether we are accurate or not.

Thus far, based on the above signs indicating the Last Hour, we can assuredly state (**and this is *our* view, among the views of many like-minded scholars**) that the Last Day, or Last Age, has already begun, and as a matter of perspective, we have identified ourselves to be currently in the Final Hours of this 'Day'.

If these Signs mark the End, what marked the Beginning of the Last Day?

Nabi Muhammad (peace and blessings be upon him), as the Seal of the Prophets, and the revelation of the Holy Qur'an, marked the Beginning of the Last Age of humanity on earth, chiefly because he was the Last Messenger, and the Qur'an was the Last Revelation. Both were among the First Signs, the Qur'an itself being described as the ultimate *Ayah* (sign). Several other signs marked the beginning of the Last Day, most of which were based on the decisive stances taken by the rejecting Jewish Doctrines of Yathrib.

Our purpose here is not to explore this subject in depth, but to understand the relevance of Time and its implication on us who will be resurrected on the Day of Judgment.

It is imperative to then identify our current states of existence in order for us to rectify ourselves, should we find that we are deviating from the rightful path. In order to do that, we need to correctly understand the reality of *what happened,* so as to understand *what is happening.* Should the modern world reveal an illusory state of existence, we who find ourselves closer to *Majma ul-Bah'rayn* become of the few who can correctly identify the deception and turn away from it.

Similar to reading a book, assuming that we are reading the 'Book of Man' depicting the entire story of Humanity from beginning to end, here we find ourselves in the final chapters of the book. The reader, hence us, will tend to read the book quicker towards the end, as compared to when we first began reading. The interesting thing to note is that as we approach the Final Chapter, we tend to slow down, depicting the slowing of Time on the Event Horizon. As we draw closer to the end, we pick up speed exponentially, hence attesting to Time moving faster toward the climax, the Singularity.

This is given by three main reasons.

One, the 'learning curve' is greater in the beginning as we steadily continue to grasp the 'norm of things'. This conforms to Knowledge in its raw form, similar to a child who is learning the world as compared to an adult who has already learned most of the world. It is difficult, and hence slower, for the child to adapt, as compared to the adult who already understands most things. Humanity, therefore, makes a slowed progress in the beginning as it tries to learn everything it discovers, progressing faster with an increase in Intellect and Knowledge.

Two, we are, by default, a hasty creature, ever determined in reaching the destination quickly. The more we learn, the swifter we advance, and are ever seeking new things to fill the void of 'knowing' within us.

Three, like every book, the crux of all the accountable and documented events are always found in the final chapters of the whole story. These are a multitude of 'closings' to every other subplot, abiding by the principle of 'everything that has a beginning has an end'. We hasten towards the end, because we are ever seeking answers to everything else that has happened, and is happening, thus far.

In order to validate all our deductions of what happened and why it happened, in order to fully grasp and comprehend why we were sent to earth and what purpose our lives served, one must continue on to the end of the Book, and it is at the very last moment, at the very end of the last chapter, that the entirety of all the plots, twists, and mysteries are uncovered.

The same can be expected in reality. It is at the very end that the truth is holistically unveiled.

The Holy Qur'an says;

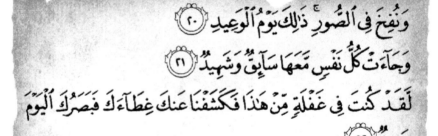

And will be blown the Trumpet; Verily, this is the Day you were warned of. [20]

And will arrive every Soul (Nafs) with it an attendant and a witness. [21]

Indeed you were heedless of this; So we have lifted your veil, so your sight Today (on the day of Qiyamah) is sharp

[Surah Qaf v. 22]

Following the Trumpet, one of the Signs of the Last Hour, the similitude of gathering is like every courthouse. Every Soul (*Nafs*) shall hence be brought forward to atone, escorted by an attendant (some scholars and commentators of the Qur'an have said that this would be an appointed Angel), and a witness who will corroborate every deed as it is judged. The most important part of all this is human perception. The *veil* that will be lifted and the *sight* which will be refreshed.

It should be noted that the 'sight' does not denote the external vision of the eyes, as we who will be summoned to Allah's court on that day will not be in the form of the *Jism* (body) we currently occupy, but a *Jism* akin to our *Nafs,* our 'self'. Hence the use of the word *Nafs* in verse 21 above, as *every-self* instead of *every-body*. The 'sight' in this case will be that of *internal* sight, a sight that brings about the elements of 'realization', 'understanding', and 'intuitive perception'.

This renewed perception will also be of Time, revealing itself in its absolute form as compared to the Mechanical form we have grown so accustomed to. This resurrection and renewed perception is similar to how the Companions of the Cave were raised, their arousal from a 'state of unconsciousness' to 'consciousness' was so that they could try and comprehend how long they had remained in that state (*Kahf v.19*). Their event was marked as a 'sign unto the people' (*Kahf v.21*). The palpability of this perception will be brought about by an alteration in what we perceive as the observable universe, as stated;

يَوْمَ تُبَدَّلُ ٱلْأَرْضُ غَيْرَ ٱلْأَرْضِ وَٱلسَّمَـٰوَٰتُ وَبَرَزُواْ لِلَّهِ ٱلْوَٰحِدِ ٱلْقَهَّارِ ﴿٤٨﴾

That Day, will be changed (transformed) the Earth, other than the Earth (into another), and (so will be) the Heavens, and they (all the creatures resurrected) will come forth before Allah, the One, the Irresistible (the Almighty, the Supreme, the Overpowering)'
[Surah Ibrahim v. 48]

It is not known how this 'transformation' will take place. Whether the Earth will be deconstructed and reconstructed anew, or an entirely new dimension will be unveiled before us, or the entirety of our existence will be transformed into another realm.

There are numerous exemplifications and explanations of the transformation of universal matter, on the Day of Judgement, all of which are defined in the Holy Qur'an and the Hadith. Some of these numerous transformations would occur on the earth in the form of earthquakes (*Surah Zalzalah*) and phenomenal geologic activity, as well as unto the Heavens (in the verse above) where even the Sun, Moon, Planets and Stars will undergo a severity of transformations.

Some have speculated over a possible disintegration into a Supermassive Black Hole (possibly one in the center of our Milky Way), which would take the entirety of what we perceive as the observable universe through a Cosmological Singularity and into 'Another World'.

Nevertheless, all these remain as speculations, and of the speculators there are those who are either doing so in mockery, and those who do not think the end will come, hence they are among those who disbelieve, ever seeking a means to alternate realities and eventual material immortality.

Of them, the Holy Qur'an says;

يَسْتَعْجِلُ بِهَا الَّذِينَ لَا يُؤْمِنُونَ بِهَا وَالَّذِينَ ءَامَنُوا مُشْفِقُونَ

مِنْهَا وَيَعْلَمُونَ أَنَّهَا الْحَقُّ أَلَا إِنَّ الَّذِينَ يُمَارُونَ فِي السَّاعَةِ لَفِي

ضَلَالٍ بَعِيدٍ ۝

They seek to hasten it (the Hour) those who do not believe

in it; And those who believe are fearful of it, for they know it is

the Truth; Unquestionably, indeed those who dispute (argue and

speculate) over the Hour are deep in error.

[Surah Ash-Shuraa v.18]

There is absolutely no doubt, that there are those who are adamantly working towards ushering in the Dajjal, hence hastening towards the Last Hour.

It is this hastening that is bringing us ever closer to the climax of existence. The verse is not only referring to *them* who are ushering the End, but also to others (even among Muslims) who are constantly disputing and propagating the end without any knowledge of it.

Refer back to the previous chapter concerning those who attempt to use calendars and Zodiac signs to 'unravel the future', hence speculating over the End. Their speculations cannot be proven, nor will they ever be proven, because Allah Almighty has Himself stated that the Knowledge of the Last Hour, hence the Knowledge of all that would occur, is known only to Him.

يَسْتَلُونَكَ عَنِ السَّاعَةِ أَيَّانَ مُرْسَٰهَا قُلْ إِنَّمَا عِلْمُهَا عِندَ رَبِّى لَا يُجَلِّيهَا لِوَقْتِهَا إِلَّا هُوَ ثَقُلَتْ فِى السَّمَٰوَٰتِ وَالْأَرْضِ لَا تَأْتِيكُمْ إِلَّا بَغْتَةً يَسْتَلُونَكَ كَأَنَّكَ حَفِىٌّ عَنْهَا قُلْ إِنَّمَا عِلْمُهَا عِندَ اللَّهِ وَلَٰكِنَّ أَكْثَرَ النَّاسِ لَا يَعْلَمُونَ ﴿١٨٧﴾

They ask thee about the Hour, 'When will it be?'; Say (unto them), 'Verily its knowledge is only with my Lord. No one can reveal its (actual) Time, except Him; It lays heavily in the Heavens and the Earth; It will not come to you but swiftly (it will come suddenly)'; They ask thee as though you are well aware, say (unto them) 'Its knowledge is only with Allah, but most of mankind is heedless.'

[Surah Al-A'araf v.187]

The gravity of this verse would take several volumes to write and explain, but to surmise its simplicity, we can understand a few things. Firstly, all Knowledge concerning the manifestation of the Last Hour exists only in the Hands of Allah Almighty. Secondly, we are drawn to the phrase *'It lays heavily in the Heavens and the Earth'* which has a variety of meanings.

According to Ibn Abbas, *'All creatures will suffer its heaviness on the Day of Resurrection'*, and according to Ibn Jurayj, *'When it commences, the Heavens will be torn, the stars will scatter all over, the sun will be wound around, the mountains will be made to pass away and all* of which Allah spoke of will occur. This is the meaning of its burden being heavy.'

Another narration and deduction, and one profoundly analyzed by Dr. Israr Ahmed, is that its 'burden' is within itself, meaning that the end of the Heavens and the Earth in the Last Hour is *in and of itself.* The similitude made is like that of the meaning of a poem within the poem, or that of death being within life, meaning we who are alive, carry our deaths with us, such that it occurs as ordained wherever in Time and Space we may be. The implication of all these interpretations is that the Last Hour is inescapable. Its knowledge of 'When' and 'How' belongs only to Allah Almighty, but its manifestation upon us (we who will feel its burden) will be as an eventuality of *our own endeavors and actions.*

This is what is meant by all events adhering to Time, with Time governing all events, and all events converging into a Singularity, a point of no return.

Given the expanse of Space, one can only envisage the length of Time endured between the beginning of this transformation and its end, and thereafter the Day of Judgment. This will further prompt a new concept of Time, unveiled before us, of which we will be questioned about.

Now we arrive at the ultimate test, which is the test of Life. Man's most precious commodity has never been land, food, shelter or clothing, but Time, and the test is an account of how that Time was spent.

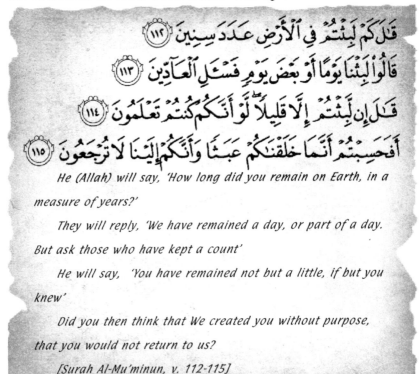

He (Allah) will say, 'How long did you remain on Earth, in a measure of years?'

They will reply, 'We have remained a day, or part of a day. But ask those who have kept a count'

He will say, 'You have remained not but a little, if but you knew'

Did you then think that We created you without purpose, that you would not return to us?

[Surah Al-Mu'minun, v. 112-115]

Time is the governing principle of life. Time is but all the currency in man's possession. From the moment of birth, we know nothing but Time, and in all its actuality, we know not how much Time we have. The Time we spend, and how we spend, will be the crux of all judgment on the Day of Judgment.

The questions now asked, are these;

If you have realized the gravity of the above verses, what is it that now draws your innermost desire? What do you now fear most? When the reality of Time is

132

unveiled before us, what will be your response to the above Divine Question?

The pagan Arabs used to regard Time as their enemy, proclaiming death at the hands of Time. This was not a notion only held by the Arabs, but by every disbeliever who strayed away. It is a statement made even today, and is widely used as the front for battling Age and Death. (*Abyss of the New World*)

وَقَالُوا مَا هِيَ إِلَّا حَيَاتُنَا الدُّنْيَا نَمُوتُ وَنَحْيَا وَمَا يُهْلِكُنَا إِلَّا الدَّهْرُ وَمَا لَهُم بِذَلِكَ مِنْ عِلْمٍ إِنْ هُمْ إِلَّا يَظُنُّونَ ﴿٢٤﴾

And they (the disbelievers) say, 'What else is there to life but this world? We shall die, and we live, and nothing but Time (Ad-Dahr) can destroy us.' And they but have no knowledge. They engage in nothing but conjecture (speculate/assume).

[Surah Al-Jathiyah v. 24]

The modern world does not recognize anything beyond the material, and if at all there is an attempt, it is sourly misguided by fantasies and myths, most of which end up in the same dark alleyways as black magic, sorcery, and witchcraft.

Time has thus become a commodity. We trade our Time for currency, per hour, per week, per wage, per salary. And when we do it on Interest, on *Riba,* it is given a fancy name like 'Overtime'.

PART TWO

THE
DIVINITY
OF
COSMOLOGY
AND
LIGHT

OF
COMMAND
AND
CREATION

An 'accident' or 'spontaneous' event is not so, except that it is a consequential effect through a series of events from a point of origin, like the final piece in a chain of dominoes all unseen but for the toppling of the last. It is not chance that something happens, except that its effect was preordained by Allah Almighty, who is well aware of the entirety of every outcome desired from His Will and Intent.

This is what is meant by the difference, as well as the integration, between The Command of Allah and the Creation of Allah, and we will explore it in further detail as we continue to study the subject of Cosmology.

In his astounding and profound works, *The Qur'anic Foundations and Structure of Muslim Society,* Dr. Fadhlur-

Rahman Ansari defines the structure of the cosmos, not as 'cause and effect' as science defines it, but as Divine Command and its corresponding manifestation of Creation. We urge the reader and researcher to dig into his voluminous works.

At the core of human existence is a 'way of life'. A set of rules and guidelines which the 'self' abides by, to progress through its age on earth. Before we even delve into Religious guidelines, at the very elemental level, a human being must adhere to three basic essentials of survival, namely Food and Water, Shelter from the environment, and Clothing from shame. (See the second-last chapter)

Each of these essentials has a set of governing guidelines, for instance, food through foraging and hunting, shelter through structure and roofing, and clothing through external skin and layer as a concealment.

As the essence of human intellect advances through the complexities of life, so too must the guidelines evolve, advance, and address every macro and micro requirement. It holds, therefore, that man, even in his most primitive state of existence, does not bear what is propagated as 'Free Will'. He possesses an ability defined as *Choice,* and Choice takes place within the umbrella of a set of rules, the above being the rules of survival, ordained by Allah Almighty.

On every level, this factual analysis challenges the secular scientific *theoretical* doctrine that man evolved from an animistic and bestial state of existence into an intellectual being.

God Almighty, the Creator of the human being, declares otherwise. Because the truth of the matter is that Adam (peace be upon him), the first man, was not a primitive, ignorant, and obtuse being. He was created with a level of intellect and granted a measure of knowledge (*Baqarah v.31*). As clear as daylight, the implication is that man did not arrive on earth with 'Free will'. He arrived on earth with a set of rules, a level of intellect, and a fair measure of knowledge.

Now, as man adhered to his environs, his intellect and knowledge would have certainly evolved, as it has over time, and so too have the rules and guidelines. In the incident between Haabil and Kaabil, the two sons of Adam, the reality of death and burying the dead was unknown to humankind, and was hence taught to us by Allah Almighty using two crows as an example (*Ma'idah v.31*).

Regardless of generation and complexity, regardless of situation and circumstance, man still only possesses the 'ability to choose'. A choice between Good and Evil (*Balad v.8-10*).

For a greater part, these guidelines have set man's 'humanity' to adhere to certain duties and morals, by which humanity itself can be described.

The questions then posed, are these;

Are these duties and morals of 'humanity', to humanity, meaningful in any way? Of war, corruption, pretense, wickedness, pollution, and overall, inhumanity itself... is 'evil' an opus of the devil, or merely his tool of deception?

Logic always dictates an 'if-this-then-that' rule, an 'either-or' guideline. If God created the Angels as a symbol of Angelic Good, then did God create the devils as a symbol of Demonic Evil?

Wrong.

Allah Almighty created both Man and Jinn with an ability to choose between serving with dutiful good and moral, or descending the steps into an oblivion of evil, where evil is not a measure of good, but a measure of the *absence* of any good, replaced by its exact opposite as the Jinn or human 'self' succumbs to a weakness of despair, temptation, lust, greed, desire, envy, arrogance, defiance, and hate.

What then is the purpose of duties and morals to the existence of man? Are they in any way meaningful?

Yes, they are. Because in His speech, the Holy Qur'an, Allah Almighty defines the essence of Creation across the Cosmos in two planes of existence, and in both planes, He has emphasized the existence of Man and Jinn. The transcendental plane, or spiritual plane, is not one of fantasy and myth, or just a 'regulative idea', but a fact— a very basic fact. The choice is upon the 'self' and the 'intellect' to understand, and have Faith in what the Creator is telling us.

The Materialist will hold a materialistic view of life and the observable universe.

The Naturalist will hold a naturalistic view of life and the observable universe.

The Spiritualist will affirm a higher state of existence in the Cosmos.

The Enlightened one, and hence, the 'self' which is at peace with itself and its existence, will find a 'balance' akin to that which the Holy Scriptures define, and this 'balanced state of existence' is the likes of which every Messenger of Allah Almighty has preached and lived (*Al-Fajr v.27*). If this is the Message, delivered from the *A'arsh* to the *Ardh,* it holds therefore, that there is unique relationship between the Creator, Allah Almighty, and His Creation, Us.

Before we delve into Qur'anic Cosmology, let us try to understand this Divine relationship between the Creator and His Creations.

In the Qur'an, Allah Almighty defines His power in relation to the cosmos in two levels, distinguished *and* integrated with each other.

These are the levels of *Al-Amr* (the Command of the Creator) and *Al-Khalq* (the manifestation of His

Creation). Both are established and united under the attributes of Allah which relate to cherishing (*Al-Wadud*), nourishing (*Ar-Razzaaq*), evolving (*Al-Baari*), and prevailing (*Al-Qahhaar*), all falling under his Supremacy as the Lord (*Al-Rabb*).

إِنَّ رَبَّكُمُ اللَّهُ الَّذِي خَلَقَ السَّمَوَاتِ وَالْأَرْضَ فِي سِتَّةِ أَيَّامٍ ثُمَّ اسْتَوَىٰ عَلَى الْعَرْشِ يُغْشِي اللَّيْلَ النَّهَارَ يَطْلُبُهُ حَثِيثًا وَالشَّمْسَ وَالْقَمَرَ وَالنُّجُومَ مُسَخَّرَاتٍ بِأَمْرِهِ أَلَا لَهُ الْخَلْقُ وَالْأَمْرُ تَبَارَكَ اللَّهُ رَبُّ الْعَالَمِينَ ﴿٥٤﴾

Verily! Your Lord (is) Allah, who created the Heavens and the Earth in Six Days (periods, ages, epochs) then He ascended on His Throne; He covers the night, the day seeking it in rapidity, and the Sun and the Moon and the Stars subjected to His Amr (command); Unquestionably, for Him is the Creation (Khalq) and the Command (Amr); Blessed be Allah, the Lord of the Worlds.

[Surah A'araf v.54]

And so, Creation first begins with His Will, His Command.

إِنَّمَا أَمْرُهُ إِذَا أَرَادَ شَيْئًا أَن يَقُولَ لَهُ كُن فَيَكُونُ ﴿٨٢﴾

By his command (Amr), when he wills something (into existence), He says to it (by exerting his command on it), 'Be' and it becomes (following His command).

[Surah Yasin v.82]

The first stage of creation begins in a stage of *Al-Amr*, or affirmed in terms of 'Be', as the Divine Directive, before the 'Becoming' of matter. We can understand this as the 'Stage of subtle existence, intangibility, or *Spaceless-Timeless*'.

If we follow the scientific definition of the creation of the observable universe, we arrive at 'evolutionary' creation beginning with the 'Primeval Particle' as the point of origin, from whence blossoms the unfolding and expanse of the observable universe through an evolutionary process of Matter, Knowledge and Time, described in the Qur'an as a 'Six Day' process.

Respective to the material and observable universe, this creative process moves from a decrease in subtlety, refinement, intangibility, and 'weightlessness', to a progressive increase in concreteness, tangibility, crystallization, and quantitativeness. From what could be described as, *'something to everything'* as compared to the secular scientific notion of 'nothing to everything'.

During this process, we find that almost everything with tangible mass and physical occupancy can trace its roots to the scientific periodic table, where every element is either an atomic unit by itself (with a composition of Protons and Electrons), or a natural composition of base elements. This is an ongoing process, which in itself is evolutionary, which then attests to Allah's attribute as the Creator. *Al-Khaaliq.*

143

The questions then posed, are these;

Does the Command only occur in the primeval point of origin?

If so, how does everything adhere to His Command?

If not, how does the Command come forth and influence every occurrence in real-time?

Following the previous eight chapters, we have now understood 'Time' to be everything more than what we physically and biologically perceive. Hence, 'real-time' bears no significant value before the Creator, Who is existent in the past, present, and future, all within the same moment. Further to this, how can we relate to a human ability of Choice as well as the Creator's attribute as *'Fa'alun Lima-Yureed'*? Doer of what He intends (*Buruj v.16*), meaning He can intervene whenever He so wills.

Surely there must exist a sublime link between He Who can do as He wills, and we who can also decide for ourselves?

The sublimity of this, is that, in the Believer's life, every Godly intervention results in rightful guidance and inspiration, and in the disbeliever's life, they may plan their plans, and Allah Almighty plans His plans (*Anfaal v.30*).

Is it not a marvel then, to uphold the human 'self' on a level of higher significance, than what the secular ideology would propagate? The human being who possesses a unique life-force called the *Ru'h*, is defined as;

وَيَسْـَٔلُونَكَ عَنِ ٱلرُّوحِ قُلِ ٱلرُّوحُ مِنْ أَمْرِ رَبِّي ... ﴿٨٥﴾

And they ask thee (O Muhammad) concerning the Ru'h (spirit), tell them, 'The Ru'h is of the Command (Amr) of my Lord...'

[Surah Isra v.85]

The essence of the human spirit is also defined as;

وَإِذْ قَالَ رَبُّكَ لِلْمَلَـٰٓئِكَةِ إِنِّي خَـٰلِقٌۢ بَشَرًا مِّن صَلْصَـٰلٍ مِّنْ حَمَإٍ مَّسْنُونٍ ﴿٢٨﴾ فَإِذَا سَوَّيْتُهُۥ وَنَفَخْتُ فِيهِ مِن رُّوحِي فَقَعُوا۟ لَهُۥ سَـٰجِدِينَ ﴿٢٩﴾

And when the Lord said unto the Angels, 'Verily! I will create a being from clay, from black mud wrought.

When I have fashioned him and breathed into him from My Spirit (Ru'h), then fall ye to him in prostration.

[Surah Al-Hijr v.28-29]

The distinction between first creating a lifeless being, *then* giving life to it, foremost defines the Prime Power of Allah Almighty as having the ability to do *both* for the purpose He intends, that is, to Create *and* Give Life.

Adhere to this distinction as it will become vital in understanding the following analysis.

One must also acknowledge a Passage of Time enduring

between the *Formation* of the being and the eventual *Breath* of the being, a Divine act which is not only a process between Command and Creation, but also intentionally so, and uniquely so *only* to Allah Almighty.

In addition to the process of creating man with a life-force from the Creator Himself, that is, of His own Spirit, it also plays a significant role in defining the essence of man with three unique characteristics, all of which are manifested in man's predestination on earth, as well as a continuous trait through humanity.

One chief characteristic is that of spirituality itself (*Fitrah*). It is the natural incline towards a higher plane of existence, be it an enlightened belief in Allah Almighty or just an emotional incline towards the presence of Allah Almighty. Such a characteristic is best observed in a child who has not yet been heavily subjected to the norms of life. This inclination to a spiritual realm, as much as modern secular ideologies will try to quash, is hardwired into man's actuality, because the *Ru'h* is always bent towards returning to the Creator.

The second chief characteristic, that of 'humanity' affiliated with a goodly nature, is evidence of this spiritual inclination, because at the core of existence, man is always seeking to do good unto himself as well as others. This inclination follows the Godly attributes of Love and Affection across every state of existence, wherein even sorrow is an attribute of love as opposed to woe and despair.

The third chief characteristic, and one that is entwined in *Amr* and *Khalq*, is given by the former verse above (*Isra v.85*), the *Ru'h* being of the *Amr* of Allah Almighty. Herein do we find the human ability to 'create' and 'innovate'.

Now, sooner will an ignorant mind jump to conclusion, condemnation, and judgment without fully realizing what the above statement means.

It does not refute that everything in the Heavens *is* of the Creation of Allah Almighty, neither does it refute that everything on Earth is *also* of the Creation of Allah Almighty. *Every single object* in existence, *animate* and *inanimate,* is of Allah Almighty's Creation, *without a shadow of doubt.*

The question then posed; is this;

If the carpenter fashioned a table with his own hands, using his own intellect and skill, how can he affirm the table to be Allah's Creation, and not his own?

The answer is very simple.

Man, who has been gifted with a Spirit from the Lord Himself, is also a vessel that *partakes in the process of Creation.*

Recall now the distinction made between Allah Almighty who can both Create and Breathe Life. The difference is that man is *limited* in his creation, Allah Almighty is *Infinite.* He has the ability to *give life* to His creation, while man *cannot* give life to his own creation.

147

Very like the mother who gives birth to a child.

The child is of her womb. It is nourished by her womb. She, who also exists by the Spirit (*Ru'h*) of her Lord, is the vessel that partakes in the process of creating the next generation. This is one such example of how the Command is ever present with every step of the Creation. It is a continuous issuance from Allah Almighty (*Dukhan v.5*), bringing every Creation into manifestation, animate and inanimate, directly or through an intermediary. Humankind, as one of these intermediaries, hence fulfills one of its roles of being a Vicegerent on Earth.

When Allah Almighty gives the Command 'Be' unto a newborn's existence, His *Khalq* (the mother) carries out the Command in a manner accorded by Him, and the process begins with the embryo in her womb until she has given birth. *She* did not create the child. She only *partook in the process* as a *vessel*.

In summary, Command and Creation are always enduring, in cosmological levels as well as infinitesimal levels, from the largest of the stars and galaxies, to the smallest of quantum particles. Every object either obeys a direct instruction from Allah Almighty, or follows the natural laws preset by Him.

Which then, of the favors and signs of your Lord will ye deny?

THE TRAVELER
AND
THE OBSERVER

Solid, liquid, and gas, the tangible compositions of the observable universe, form the entirety of all that is observable from particle to planet to star. The process of formation from either state to the other must abide by the rules of Time, as well as elements of Temperature and Pressure which must also exist for any transformation to occur, and every transformation must undergo its respective process of deconstruction from one state of existence to reconstruction into another. Fundamentally, no transformation can occur without undergoing a process relative to the motion of Time moving in a forward direction.

At the Time of observation, the observable universe does not reveal itself in a 'present' state, rather it reveals a state as it was, variant to the Time taken for visual information to travel from the point of origin to the destination. Sight cannot function without light reflecting or emanating from the observable object.

Similarly, the observation of stars and planets are only possible because of the Light and Energy they emit or reflect in the earth's direction. Again we see the concept of Time, because even the journey of Light, however fast it may be, must adhere to a Passage of Time over Distance as how much Time is spent for every bit of information in every stage of its travel to arrive at its destination.

What we see of the stars is a package of luminous information sent eons back, arriving on earth as it were whence it first left its point of origin. Hence the measurement of distance in 'light years' as the time taken for 'light' to travel in a year against its speed. The calculation of a Light Year is as follows;

Distance *(light year) is given by*

The *Speed of Light*

multiplied by

Time *taken to travel (one earthly year)*

The variables used are;

Speed of Light *in Km/s is 299,792 km/s*

Time *in seconds is 365 days x 24 hrs x 60 mins x 60 secs*

= 31,536,000 seconds

150

Therefore;

One Light Year = 299,792 x 31,563,000 =

$$9.46 \times 10^6$$

or

9.5 Trillion Kilometers

as the distance that Light can travel in one earth year.

By the above illustration, we can begin to appreciate the vastness of space and the expanse of the observable universe.

As we have identified, that because Time exists in a form of intangibility, for the essence of its comprehension on earth, we have developed strategic tools for measuring its passage in relation to the motion of the earth around the sun. Here we are able to distinguish between the tangibility and intangibility of Time. In its earthly tangible context, it appears finite, but in the Cosmos, it is intangibly infinite. Similarly, its intangible Knowledge is intangibly infinite. Because we cannot palpably relate to the entirety of the Cosmos, Allah Almighty bestows its Knowledge unto us in tangibly documented forms, words, phrases, and signs revealed in the Holy Qur'an, in addition to numbers, patterns, and shapes.

The study of the observable universe, given by its expanse, is complex for two main reasons. We, the observer, the students, have to formulate our study while being grounded to the earth, and so the entirety of the study is made only through sensory outreach, namely sight and sound. To extend our sensory reaches, through Telescopy and Instrumentation, Science and Technology have enabled the mapping of the universe along the invisible ranges extending beyond the bandwidths of human visualization along the electromagnetic spectrum. Background radiation and microwave mapping have allowed the intellectual astrophysicist to trace back the origin of the observable universe and calculate its age by a measure of earthly years. Instrumentation has also enabled a near-accurate understanding of the formation of the observable universe as far as planets and stars are concerned, by way of measuring the emittance of light and radiation.

The complexity of the study is such that, by and large, every aspect of the observable universe that we have come to understand through science is purely theoretical, as much as learned scientists and academics would choose to argue. We can boldly affirm this claim by what the Qur'an tells us;

يَـٰمَعْشَرَ ٱلْجِنِّ وَٱلْإِنسِ إِنِ ٱسْتَطَعْتُمْ أَن تَنفُذُوا۟ مِنْ أَقْطَارِ ٱلسَّمَـٰوَٰتِ وَٱلْأَرْضِ فَٱنفُذُوا۟ لَا تَنفُذُونَ إِلَّا بِسُلْطَـٰنٍ ﴿٣٣﴾

O ye of the Jinn and Men, if you (think) you have the ability to venture beyond the frontiers of the Heavens and (penetrate the depths) of the earth, then proceed; Ye will never venture forth, save without Our authority.

[Surah Rahman v.33]

152

This verse requires little in terms of interpretation to understand its explicit definition and challenge contained, a challenge which would require an immeasurable amount of audacity from Jinn and Man to stand up to. The verse also defines an unspoken statement in its finality, that Cosmological Journey *is possible* for our kinds, but with Allah Almighty's permission.

Historically, three (that we know of), out of the entirety of human existence, have ever ventured across the Heavens without the need to invent life-supporting tools to enable man's elevation even a few feet off the ground. We are not referring to the numerous astronauts and cosmonauts who have traversed as far as the moon, nor the unmanned crafts sent to Mars and as far away as interstellar space. As much as their efforts and achievements can be commended and applauded, they have merely taken an infantile step forward thus far.

We are speaking of those individuals who have traversed the entirety of the Cosmos, without the need for *any* technology. Their journeys are not only astounding and marvelous in and of themselves, but sadly, much of the world chooses to commemorate the modern man's flight to the moon with a greater zeal and enthusiasm, rather than enlighten themselves with an actual venture through the Cosmos by these three unique and blessed individuals.

Nabi Adam (peace be upon him) was the first man to descend to earth from the Heavens, a journey well across the vastness of the Cosmos;

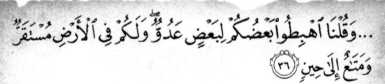

...وَقُلْنَا اهْبِطُوا بَعْضُكُمْ لِبَعْضٍ عَدُوٌّ وَلَكُمْ فِى الْأَرْضِ مُسْتَقَرٌّ وَمَتَاعٌ إِلَى حِينٍ ﴿٣٦﴾

...and We said (unto Adam); Descend (from the Heavens), of

you (Mankind) and of them (Shayateen) enemies of each other; and

for you on the earth is a dwelling place and provision for a period

(until the Day of Judgment)

[Surah Baqarah v.36]

Following the injustice and wickedness enacted by the Pharisic Rabbis, resulting in a series of persecutions, Nabi Isa (peace be upon him) was raised up to the Heavens, a journey soon to be completed with his nearing return back to earth;

﴿٥٥﴾ ... إِذْ قَالَ اللَّهُ يَا عِيسَى إِنِّى مُتَوَفِّيكَ وَرَافِعُكَ إِلَىَّ

And when Allah said unto Isa son of Maryam (Jesus son of

Mary), 'Verily I will take you (away from this earth) and raise you

unto Myself (in the Heavens)...

[Surah AL-Imran v.55]

At a time in his life, when Nabi Muhammad (peace and blessings be upon him) was faced with the grief of losing both his beloved uncle and first wife, Khadija (peace be upon her), the Holy Prophet was taken on a nightly journey so miraculous, no human in history had ever merited such honor. The event began with a journey across the plains of the earth;

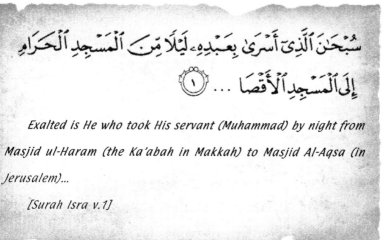

Exalted is He who took His servant (Muhammad) by night from Masjid ul-Haram (the Ka'abah in Makkah) to Masjid Al-Aqsa (in Jerusalem)...

[Surah Isra v.1]

The journey endured into the Heavens from that point, described in intricate detail in the verses of Surah Najm and the Books of Hadith.

His journey, by far, is the most astounding of all three, not only traversing an earthy distance quicker than light itself (although science would elaborate that nothing can travel faster than light), but crossing the physical barrier of the observable universe into the rest of the Cosmos, arising and experiencing events and occurrences within each Heaven until he arrived at the *A'arsh* of Allah Almighty.

To any scientist, any physicist, astronaut or cosmonaut, or any ordinary man for that matter, *this* should be the pinnacle of *every* Cosmological study, for where else could we have possibly learned of what transpires in the Heavens above, but from the experiences of the Blessed Prophet himself.

So miraculous is this journey that not only would he be the first man to have traversed the Cosmos from earth and back (the second being Nabi Isa who is yet to complete his journey), but the only man to have had the highest honor and esteem of leading *Salaat* in the company of the greatest men to have walked the earth, all the Blessed Prophets of Allah Almighty.

The entirety of this miraculous journey is so profound and intricate, it would take pages over pages to speak of it. For this reason, we have dedicated a separate study and publication for the future, *Insha'Allah.*

These three blessed individuals, out of the entirety of mankind, were the great travelers of the cosmos. We who have not merited such elevation, remain as observers, and thus do we observe.

When we look into the night sky, when we stare into the glimmer of the stars, we are, in fact, looking back in Time, and yet when we travel into the night sky we are traveling forward in Time. What we see, grounded to the earth, is pockets of information delivered along luminous spectrums instead of words, far removed from the conventional assumption of what we perceive as 'information', but in essence that is what we are receiving. The same notion exists within our earthly confines as well, but given the shortened distances of communication, in contrast to the vastness of space, the effect of Time Loss and Dilation are not so easily noticeable.

The questions then posed, are these;

How far back in Time and Space can we really see? With a big enough telescope, could we see the first stars and galaxies in formation? Could we possibly see the actual formations taking place, albeit the duration of formation would far supersede any man's existence? Look far enough, and could we perhaps see the point of origin itself, the particle from whence it all began?

The expansive study of the Big Bang theory states that the formation of the observable universe occurred in stages, or eras, each one with a specific purpose and each one enduring a specific period. This theory is basically a History of Space-Eras. For instance, there was an era when 'energy and matter' took formation, an era where 'energy and matter' were coupled, then separated, an era when atomic structure took form, an era of ionization, and so on until the era of the observable universe's population with galaxies, stars, and planets.

Even though each era endured through a sequence of events, given the vastness of space and the duration of information transmitted through Space, every one of these eras is still theoretically observable. In order to achieve this feat, our eyes must look through seven historical eras, namely the Solar System Era, the Galaxies Era, the Protogalaxies Era, the Re-ionization Era, the First Stars Era, the Dark Ages Era, and finally the Flash of the Big Bang.

Conclusively, however, this only gives us the limited knowledge of the observable universe. In the context of this study, our objective is not identify what science has already been trying to identify over the centuries. Our objective is to understand the *Cosmos* as Allah Almighty has explained to us, and for that purpose, we the observers, need not look through the telescope, but through our inner eye.

THE
COSMIC
CALENDAR

By definition, History is the study of past events which have made a mark on a particular Timeline. While the mainstream outlook of History only pertains holistic political, geopolitical, or societal events, the essence of History includes every individually significant event as well. Biographies and Autobiographies of notable entities are historic in their own respect, so too are micro-societal histories within families, clans, tribes, cultures, ethnicities, and races, eventually progressing into Nations, Kingdoms, and Civilizations.

Historical facts and events are documented based on their significance and authenticity, given by how and why these events would have played a vital role in altering or determining a particularly significant outcome. In most cases, however, certain events have a tendency to be overlooked, because of a missing link to a particular eventuality, or they are subjectively granted to be insignificant without much consideration, often leading to a falsification or inaccurate portrayal of History. In the case of Islam, for example, much of the western world documents Islamic history in negative outlooks, either deliberately out of malice, or chiefly because they tend to overlook crucial elements and events which would otherwise alter their outlook with greater authenticity. This approach, or methodology of study should be frowned upon, because it is ever inclined towards biasness, prejudice, and often jumping to conclusions. It does not only pertain to Islam, used here as an example, but to nearly every aspect of human history, writing and rewriting fact to suit a particular doctrine.

A more accurate approach to History and its study, is to regard every event as significant, whether or not it requires elaboration or mere acknowledgment. Every event, large or small, holistic or individual. This analysis of the past, with an emphasis on authenticity, is philosophically known as *Historicity*.

Individually, every moment in life is marked on a Timeline. Every moment of our lives is a decisive moment, every decision guiding each one of us through a particular path to a particular outcome, with every outcome representing itself as a decisive moment. Every decisive outcome further adds to what can be described as a 'stream' of outcomes, and as they become holistic, they eventually lead to global events and outcomes which structure or restructure, define or redefine, alter or shape, the actuality of mankind in a multitude of variable pathways, often identified as an infinite number of possible and probable routes, yet somehow all converging to a unified destination. This forward analysis is simply known as *Futurism*.

The definition of history being a study of past events can also be regarded as a documentation through memory, and in any instance of looking back into our lives, we may be accessing memories, but in essence, we are traveling back through Time. The visualization or enactments of scenes within our minds are not illusory or surreal. They are absolute occurrences, hence navigating through memories is, in essence, a journey through Time.

In similar context, looking into the night sky is a journey through Time for the observer perceiving a canvas of information delivered through luminous pockets as they travel forth from their points of origin, a planet, star, or Galaxy.

As we observe each of the eras of formation, the realization dawns on us that each era represents a unique set of events that took place during the history of the observable universe. The point of origin, scientifically defined as the Big Bang, caused an expansion such that every component is 'moving away' from each other, not just through Space, but also through Time. The younger components are perceived as moving away at incredible speeds, and because of the Time it takes for light to travel, delaying the rate at which information is delivered to the observer on earth, the further away we look into deep space, the further back in Time we look, revealing the observable universe in its youth, and further back in its infantile state.

The marvel of this phenomenon is that regardless of Time's forward motion, we who perceive it as such are actually looking at Time and its events in both the past and present state within every instance. The aspect of Time which is kept concealed from us falls under the Knowledge of the Unseen, *'Ilm ul-Ghayb*, only known to Allah Almighty. The various eras of the creation of the Cosmos are described in the Qur'an in the most eloquent of languages, words, and phrases known to man, verses that cannot be understood without the application of a proper methodology.

The inconsequence of humans is such that the entirety of our existence, with relevance to Time, is but a speck of dust in contrast to the Cosmos. Above all things, Time is the only definitive relative to human existence, yet our perception of Time often tends to become narrow-minded.

It is an immensely difficult task for the human brain to gain an accurate perception of the Age of the Observable Universe. In terms of numbers and mathematics, sure, we can draw up equations all day long, but we find ourselves lacking in scale, perception, and depth. To put matters into perspective, the difference between One million seconds and One billion seconds is 31 years, and by fractional scales, a million seconds is just about two weeks.

To a materialistic perception, two weeks may as well be discarded unless some material benefit can be drawn from that time period, but in the observable universe, the earth makes fourteen complete rotations at an immense speed of nearly 30 kilometers every second, having traversed a distance of 30 million kilometers within a fortnight.

The best scientific research estimates that the observable universe is 13.8 billion years old in earthly-time, but that number is so big, and as mesmerizing as it may appear, it means little to the ordinary man who can only conceptualize a life-span of approximately 80 years, which in and of itself is an indefinite number.

To assist the non-physicist's mind to appreciate the immensity of the observable universe in a comprehensible scale, in 1977, astronomer Carl Sagan introduced a Cosmic Calendar to the world of physics.

The Scientific Cosmic Calendar is a scale in which the 13.8 billion year lifetime of the universe is scale-mapped onto a single earthly-year. The visualization of this calendar assists the human mind to put into perspective the History of the Observable Universe into comprehensible scales. In this scale, the Big Bang took place on 'January 1st at midnight', and the current, present time is scale-mapped to 'December 31st at midnight'. The scale identifies 437.5 astronomical years into 1 earthly second, and the concept is widely accepted in the world of physics as a way to conceptualize the vast amounts of time in the history of the observable universe.

The following are a few highlights derived from this calendar. Note that these are only approximations, as the events shift constantly with every passing second.

January 1st — 14 seconds after midnight;
The observable universe is in a hot, dense and gaseous state with the formation of Hydrogen and Helium. Electrons combine with protons to form atoms. In the 13 seconds prior, it is in a plasmic state of disjointed protons, electrons, and baryons.

January 10th — 300 million years past;
The first stars ignite in their infantile states.

January 13th;
Small galaxies begin to take shape as clusters of stars.

164

March 15th;
The Milky Way takes form

End of August, beginning of September;
The Solar System takes form

14th December;
Small animals

20th December;
Land plantation and vegetation

25th December;
Hail the magnificent dinosaurs

26th December;
Mammals appear

27th December;
Birds and insects

30th December;
Extinction of dinosaurs

14 seconds ago;
Civilization.
Every person we know or have heard of, present or historical

5 seconds ago;
Nabi Isa is born unto the Israelites

4 seconds ago;
Nabi Muhammad receives revelation in Cave Hira.

1 second ago;
The Modern Age

present lives, in the grand scheme of the

e universe, is but a fraction of a fraction of

materially insignificant, according to the

calender and the expanse of Space.

The Holy Qur'an also avails a Cosmic Calendar for human comprehension, but before we draw the knowledge of the Qur'an, we must acknowledge a very crucial point.

We are not using the Qur'an to validate scientific discovery, and Allah Almighty forbid that we should ever use science to prove the validity of the Qur'an. Modern science may have worked wonders in bringing certain information to light that was not before known to man, but modern science does not govern the existence of the Heavens and the Earth. Science would claim that the great machine of the universe is governed by the laws of physics, but the believer must acknowledge that the language of science was written by man, and the governance of the Cosmos is decreed by Allah Almighty.

Therefore, with all due respect to modern science, every study in this book has been done with delicate care, placing the Holy Qur'an above all else. We will adhere to the Divine terminologies and phrases of the Qur'an, such as *Samawat* and *Ardh* in senses which are far more broader than the physical, biological, and chemical compositions of the known and observable

Earth and Universe.

The Holy Qur'an declares;

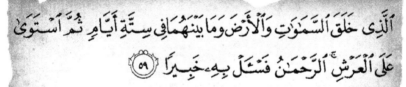

ٱلَّذِى خَلَقَ ٱلسَّمَٰوَٰتِ وَٱلْأَرْضَ وَمَا بَيْنَهُمَا فِى سِتَّةِ أَيَّامٍ ثُمَّ ٱسْتَوَىٰ عَلَى ٱلْعَرْشِ ٱلرَّحْمَٰنُ فَسْـَٔلْ بِهِۦ خَبِيرًا ۝

He who Created the Heavens and the Earth, and all that is between them, in Six Days (eras or epochs), then He established His Throne (above all). The most Beneficent! So ask of Him, He knows all.

[Surah AL-Furqaan v. 59]

This calendar depicts the Six Era's of Creation following an era (totaling seven eras) of a state of *Timelessness-Spacelessness,* or a state of pre-existence which is an era by itself. To put this into perspective the eras of man can be regarded as;

Biological Inexistence, Embryological Existence, Birth, Infancy, Childhood, Adulthood, Death, Resurrection, Atonement, and Hereafter. Outside biological existence, the pre-embryological states, as well as death, and after death, also constitute the Eras of human existence. Many scholars have further classified all these major eras into smaller, more definitive stages of actuality.

It should also be noted, from the verse above, that in its concise and explicit definition, the entirety of the Seven Heavens, including the physical 'universe' as we

167

have termed it, were all created simultaneously.

The totality of this Six Day calender narrows down the various eras of creation depicted in the Qur'an, as follows;

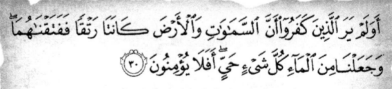

Do they not see, those who disbelieve, that the Heavens and the Earth were a joined entity, then we parted them...

[Surah Anbiyah v.30]

In the context of the verse, the phrase *'do they not see'* does not only denote a 'visual' or 'observational' connotation, which we abide by when studying the observable universe, but has a greater emphases on 'do they not *understand?'* or 'do they not *realize?'*

The emphases on 'understanding' as an attribute of the human *Nafs,* denotes more than the 'observable' universe which we have become accustomed to with our visual senses. As compared to computing rational calculations in the brain, which is a 'physical' attribute of acquiring information, the Qur'an is directing our attention to an aspect of 'belief' over rationality, with 'understanding' being the aspect within the human *Nafs* responsible for converting information into knowledge.

Understand what?

That the Heavens and the Earth, the entirety of the Cosmos, including its physical composition, were all at one point, a singular entity. The word *Ratqan* (enjoined) has an emphasis on a *'multitude of objects condensed or compressed'* into one.

Secular Science elaborately describes this singularity as a point from which one particle, wickedly termed as the 'god particle', was split apart, resulting in the expansion of the observable universe.

Islamic Science, however, presents an alternative, more accurate description. The relevance of the above word *(Ratqan)*, with its roots in *Ra-Ta-Qa* (enclose), is in contrast with its antonymic connotation, the next word, *Fafataqna,* from its root *Fa-Ta-Qa* (disjoin). In the case of the observable universe, both words are eloquently describing the physical processes of Nuclear Fusion (to fuse together) and Nuclear Fission (a splitting).

In the context of the physical universe, the word Nuclear is used because it involves the process of enacting a destabilization within the nucleus of a particle, deep within its atomic structure.

Scientifically, the Big Bang of the physical universe occurred as a result of Nuclear Fission, the splitting of a singular particle with an immense burst of energy and matter leading to the subsequent formation of all else. Science is unable to explain the origin of this point, or what they classify as the primeval particle, but the verse explains its origin as something that was 'first put together' (*Ratqan* - enjoined) as the point of creation within an era of pre-existence, as the *'Particle Created'* by Allah Almighty. We urge you to take a moment here to ponder over the vitality of this statement in accordance with what the verse is describing.

In simplistic terms, Allah Almighty *Created* the *Particle* by fusing all the required elements together, hence He *Created* the Singularity, thereafter He *split* the *Particle* (Fission) and *Created* the physical and observable universe as we have come to know it, and all the Seven Heavens alongside. He who places his faith and trust in Allah Almighty will be able to see further than a scientific explanation, that aside from matter emerging from this point of origin, so too did other planes of existence. Some have denoted the observable universe to be one of the Seven Heavens, meaning Six Heavens plus the observable universe, while others state that the Seven Heavens are seven plus the observable universe making them Eight in total. We cannot provide a definite response to either argument, thus we affirm that only Allah Almighty knows best, leaving the speculation to the speculators.

Within the following analysis, we will focus on the physical universe with regards to the Qur'anic cosmic calendar, while making educated references to the scientific calendar. Since the Qur'an does not explicitly refer to the Six Day process as 'calendar', we will hence refer to the scientific calender as Sagan's Calendar, and the Qur'anic description as a Divine Process.

According to Sagan's Calendar, the creation of the earth would have begun approximately around late 'August' with the formation of the Solar System, which would place it within the last third of the calendar.

Similarly, the Holy Qur'an has also allegorically depicted the Divine Process by setting out the Creation of the Universe in Six days, or eras, with the earth's creation taking place in the last two eras, or in the final third of the Six Day Process.

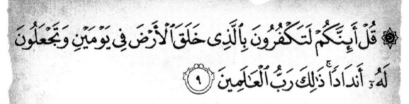

Say (unto them), 'Do you indeed disbelieve in the One who Created the earth in two days (to periods or eras)? And you bring against Him your own gods as rivals? (Know) that He (Allah) is the Lord of the Worlds.'

[Surah Fussilat v.9]

Now, according to a holistic of things, the creation of the earth could not be possible without the formation of the Solar System, hence without the formation of the Milky-Way, and by extension the formation of the observable universe. This statement, in and of itself, is quite obvious to any logical mind, the fact being that man could not exist without the earth, there being no other planet perfectly suitable (that we know of) for man to survive on.

This particular point has a force of gravity so immense, it is crucial for us to spend some time pondering the following verses;

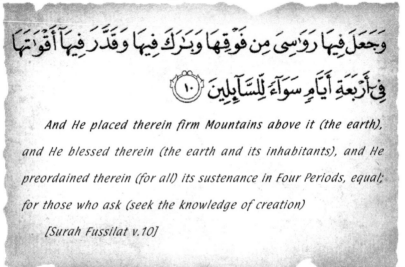

And He placed therein firm Mountains above it (the earth), and He blessed therein (the earth and its inhabitants), and He preordained therein (for all) its sustenance in Four Periods, equal; for those who ask (seek the knowledge of creation)

[Surah Fussilat v.10]

The phrase *'sustenance in four periods'* is not depicting four ages *after* the earth's creation, but four ages *prior* to its creation, the rooted indication here being the word *Qaddara*, from the root word *Qadr*, denoting the Divine Decree of Allah Almighty as a preordainment. It means that the earth and its sustenance was predetermined in the four stages *prior* to its creation. The formation of the earth could not have happened without the formation of the universe, which in and of itself is a Divine Decree of Allah Almighty. It is within this context that we place our belief on the Sixth Pillar of *Imaan,* that man's creation following earth's creation was preordained long before even the formation of the Heavens and the Earth.

Here, within this verse, we also find the initial two-thirds of the Divine Period, the first four eras, which were the determinant factors in the formation of the earth. As stars were clustered into galaxies and planetary systems were formed within the galaxies, the first four periods would have been the determinants of the last two periods when the solar system and the earth took form.

It should be noted that the verse is only describing the *earth's* formation to *our* context, whereby the reality would have seen the formation of other planetary systems across the observable universe taking place within the same, or around the same period.

It should also be noted that with regards to other possible creations in the observable universe, aside from Man and Jinn in the solar system (Allah Almighty knows best), other beings may have had their worlds created in accordance to a different segment of the process, but still within the same overall Six Day period. This thought allows for the fact that not all planetary systems were formed in the final third of the whole process. Some systems have also been known to have taken form in the middle third or even the first third, some which are

173

still taking form, and some which may still take form, and Allah Almighty truly knows best.

According to Sagan's calendar, the first 14 or 13 seconds following the Big Bang was but a gaseous and plasmic state of existence. This is indeed true as we will see below, and even though science does explain the formation of Hydrogen and Helium (first two elements on the Periodic Table) from a Plasma of Quantum Particles, we have found that the Holy Qur'an explains it better;

ثُمَّ ٱسْتَوَىٰ إِلَى ٱلسَّمَآءِ وَهِيَ دُخَانٌ فَقَالَ لَهَا وَلِلْأَرْضِ ٱئْتِيَا طَوْعًا أَوْ كَرْهًا قَالَتَآ أَتَيْنَا طَآئِعِينَ ﴿١١﴾

Then He directed Himself towards the Heaven while it was still smoke (plasmic and gaseous); He said unto it (the state of Heavens in its infancy), and unto the earth (existing here in a Timeless-Spaceless state) 'Come, both of you, willingly, or unwillingly (be obedient, or be forced to obey)', they said, 'we come willingly'
[Surah Fussilat v.11]

This verse is not following the Creation of the earth in the final third of the process, rather it is following the initial stages of creation, or it is an occurrence within the first Four eras of creation. Beginning with the first era while the Heavens were infantile, existing in a plasmic and gaseous state (smoke), and the earth was all but a Decree (*Amr*) not yet manifested into actuality (*Khalq*). The verse is also describing something unique and amazing.

Just as well as Allah Almighty made a covenant (*Al-A'araf v.172*) with every single *Nafs* (mankind in its true form) prior to the biological creation of humans on earth, so too does He make a covenant with the Heavens and the Earth in their pre-conceived state of existence. Their compliance of 'coming willingly' could be an affirmation and testimony of Allah Almighty as their creator, and Allah truly knows best.

فَقَضَىٰهُنَّ سَبْعَ سَمَٰوَاتٍ فِى يَوْمَيْنِ وَأَوْحَىٰ فِى كُلِّ سَمَآءٍ أَمْرَهَا وَزَيَّنَّا ٱلسَّمَآءَ ٱلدُّنْيَا بِمَصَٰبِيحَ وَحِفْظًا ذَٰلِكَ تَقْدِيرُ ٱلْعَزِيزِ ٱلْعَلِيمِ ۝

Thereafter He completed them as Seven Heavens in two days, and He revealed in each Heaven its affair; And We adorned the Heaven nearest with lamps and (as a) guard; That is the Decree of the Almighty, the All-knower.

[Surah Fussilat v.12]

The above mentioned two days constitute the first-third of the epoch of Creation, an era which saw a transformation from gaseous states, to the ignition of Stars and formation of galaxies. According to Sagan's calendar, the galaxies would have taken form in late March— beginning April, which would have been in the middle third of the Six Day epoch of creation.

All the above unveil a deeply phenomenal thought, that while we humans are categorical in our definitions of 'what is alive and significant' and 'what is not', to Allah Almighty, *all* His Creations have a life of their own, all bearing a significant attachment to Him. As a tendency we hardly seem concerned or appreciative to the earth we live on, the air we breath, and the water we drink, whereas to the Creator, all these elements have a breath and emotion of their own.

On the surface we display an abhorrence to polluting and extorting our environments, but we hardly seem bothered when we bore the ground to erect our buildings and houses, all the while polluting the air and water with gaseous and chemical discharges. Hardly any of us bats an eyelash when plants are sprayed with toxics, or when animals are caged behind fences.

This is not an attempt at lobbying for the environment. Just a thought to consider with regards to the above verses and the gravity of what it means when God Almighty is speaking directly to the Heavens and the Earth, and they are all submitting to His Command willingly.

THE
HEAVENS
AND
THE EARTH

Thus far we have identified the chief component of creation, Time, the documentation of creation, Knowledge, and the entirety of the Six Day process. We have correctly identified the Point of Origin and what existed prior to the origin. We have understood the sublime relation between the Creator and His Creation, and we have understood the correlation between His Command and its manifestation.

You may have noticed that throughout the book, we have continuously used the phrase 'observable universe' instead of just 'universe' by itself, and there is a reason as to why this has been deliberated. Progress from hereon bears less and less material evidence, and requires a greater advancement in Faith and Understanding. We now urge the reader and researcher to open their minds and hearts wider in order to grasp what now lays utterly beyond physical comprehensions.

Throughout history, the Cosmos has played an influential role in mankind's intellectual development. Theologically, every known, and even unknown civilization has attempted to make some connection with the worlds above earthly confines. Strangely so, it is only with the modern age that this connection has been reduced to a materialistic outlook.

Allah Almighty has told us that there has not been a nation, or a people on earth, to whom a Messenger, and thus Revelation, was not sent (*An-Nahl v.36*), and part of every Revelation, in some form or another, bears a moderate measure of the Knowledge of the Cosmos, just as the Holy Qur'an is unto us.

Albeit most revelations were distorted by man's hand, we can conclusively say that the knowledge was not. In some from or another, a pattern always emerges, one that can be likened to what the Qur'an says.

Sumerian texts reveal *'an-imin-bi ki-imin-bi'* (the heavens are seven, the earths are seven). In Judaism, according to the Talmud, the universe is made of seven heavens (*Shamayim* in Hebrew). In Christianic Apocryphal text, the Second Book of Enoch vividly describes Seven Heavens. In Hinduism, according to the Brahmanda Purana, there are fourteen worlds of which Seven above are considered Heavenly.

The above connotations are not meant to endorse any other belief system other than what the Holy Qur'an can irrefutably validate, but to provide a perspective, and hence a profound insight, that the Seven Heavens are not sensory or informative elements, rather their actualities are deeply embedded into humanity's spiritual existence, and we will explore further along what this means.

In Islamic Cosmology, one of the subjects that has always intrigued scholars has been the significance of 'Seven' Heavens, and whether it means Seven Universes, or that the number seven is in reference to an 'infinite' number of universes.

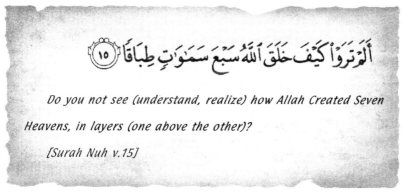

Do you not see (understand, realize) how Allah Created Seven Heavens, in layers (one above the other)?
[Surah Nuh v.15]

As noted above, most Islamic Scholars and commentators of the Holy Qur'an have used a linguistic, rather than literal approach to deciphering the significance of 'Seven' as a number.

Foremost, the word *Samawat* has widely been mistranslated as 'universes' which leads to some scholars arguing that 'seven' is not an exact count, but a connotation to 'many in number', thereby signifying that the 'universes' are infinite in number.

This assumption has been made in accordance to Allah as the Ultimate and Everlasting Creator, who continues to Create in great numbers, rather than to assume His creation to be finite to Seven universes.

Similarly, another verse says;

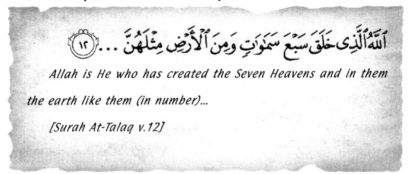

Allah is He who has created the Seven Heavens and in them the earth like them (in number)...
[Surah At-Talaq v.12]

A surface interpretation would reveal Seven Heavens and Seven Earths (similitude in number), and given the explanation above, an infinite number of both might seem more befitting, both to the essence of creation, and to Allah Almighty as the Greatest Creator. This theory would also seem to agree with the scientific hypotheses of a 'multi-verse'. As we previously identified, however, this creates a problem, as an infinite number of 'Earths' would denote an infinite replication of humanity, which is neither mentioned nor endorsed by the Holy Qur'an. Furthermore, even if this were true somehow, and Allah Almighty knows best, it would not be able to explain the relevance of an infinite number of Earths, nor would it bear any significant knowledge to mankind who is limited to one observable universe, and Islam continues to emphasize a dire need to pursue an understanding the Heavens and the Earth. In addition to this, Nabi Muhammad's nightly journey through the cosmos was definitive to Seven ascensions, thereby refuting any notion of an infinite connotation to the number Seven when describing the Heavens.

There is an alternative explanation of the Heavens being Seven in number, and one that we find bears more sense and accuracy.

A more complex analysis delivers 'Seven' as a number of dimensions, or planes of existence, within the entirety of the Cosmos relevant to mankind, the lowest of them, or the lowest below the Seven, being the quantitative, visual, and *observable universe.*

The word *Samawat* is directly translated as Heavens in Classical Arabic, because terms like 'space' and 'universe' were not only unheard of then, but were insignificant to that age, and yet the Heavens have always been significant to every age of mankind. Therefore, it holds that 'space' itself, and its physical exploration is but a small endeavor in the holistic actuality of humanity, as compared to a *transcendence through the Heavens.*

This thought is difficult to comprehend because every affiliation and perception with the Cosmos has always been made with a physical and materialistic approach, hence the reason why we urged the reader to progress hereon with a wider outlook and a strong reliance on *Imaan* to understand what it all means.

The word *Samawat* can also be translated as *Stratas*, or states of existence within the Cosmos. If we simply replace the definition of *Samawat* from *universes* to *Stratas*, a clearer interpretation of the Qur'anic verses arises. By this, we can also understand the observable universe to be 'a strata' along with 'Seven other Stratas'. We can then relate with the physical Earth as a location within a physical and naturalistic composition which is the observable universe, and seven additional planes of existence all layered upon each other. Thus, when the verses say *Sab'a Samawat wal-Ardh,* they can be interpreted as Seven Heavens and the Observable Universe, with the word *Ardh* denoting the physicality of the worlds within the context of the respective verse. Thereafter, we can then affirm that the Cosmos relevant to our existence, revealed to us in the Holy Qur'an, is but one in an *infinite number of other such creations* of which we have no knowledge revealed to us.

Theoretically and practically, this definition neither negates Allah Almighty's attribute as the Everlasting and Infinite Creator, nor does it denote an alternative to the word 'seven' in every verse. It simply denotes, that within *our* relevant Cosmos, there exist Seven varied states of existence other than that of matter and energy, stratas of higher dimensions, sublime and divine in existence.

From this standpoint, several other associated verses and Hadith become simpler to understand, including the journey of Nabi Muhammad through the Seven Heavens.

Herein do we enter the realm of inter-dimensionality in Islamic Cosmology.

It should be noted that what secular science explains as dimensionality has no significant value before the Divinity of the Holy Qur'an, because as we have already identified, the scientific rationale of quantitative analysis cannot be applied to sublime planes of existence. Not even within earthly dimensions, such as that of the Jinn, can science provide any rational explanation. Higher dimensions beyond the earth require spiritual ascension.

To understand dimensionality, we have to begin with the actuality of human existence. Some of the great scholars of Islam have deciphered human existence not only as three-dimensional objects in three-dimensional space, but in three overlayed and interlinked compositions, by separating man into a trio of components.

9th-century Islamic scholar Imam Al-Ghazzali, in his volumes of *Ihya,* defines the Body (*Jism*), Soul (*Nafs*) and Spirit (*Ru'h*) as the harmonious composition of the human being. He defines intellect, or the mind (*Aql*) as the integration of the *Jism* and *Nafs*, and emotion, or the heart (*Qalb*) as the integration of the *Nafs* and *Ru'h*. Refer back to the chapter *Of Religion and Science*.

The following study is complex and difficult to conceptualize, because it eludes any manner of rational or logical arguments. We again ask that the reader and researcher bear patience and open-mindedness as they unravel the mesmerizing descriptions of inter-dimensional existence from the Qur'an. We also ask that every intellectual mind bear witness that Allah, and Allah alone knows all concerning the Heavens and the Earth, and all that is between them.

As we have identified, the central point is not to look at 'Heavens' as physical locations of matter. The visualizations of the Heavens is but on a plane unseen to the human eye, and therefore unobservable from a scientific standpoint. The Seven Heavens cannot be measured, they cannot be quantified, documented, nor can they be studied, at least not from the physical locales of this earth nor the observable universe. It matters not where the academic has established his desk and chair, on Earth, the Moon, Mars, a habitable planet in the Andromeda Galaxy or even at the very edge of the physical universe, with his naked eye, or even with the most advanced optical technology, he will not be able to observe the Heavens. Verse 33 of Surah Rahman is very explicit in defining the Heavens as an impenetrable frontier for both man and Jinn, save with the Authority of He who Reigns above all.

Understanding the actualities of the Heavens first and foremost requires spiritual patience and a deepening adherence to *Imaan*. While this may seem irrelevant, it bears great importance in providing the 'self' with the strength to affirm one's significance, not as man, but as human and as a vicegerent of Allah Almighty. This ascension through Faith and Knowledge provides one with an inner understanding of the Heavens with the use of internal sight, derived from significant and virtuous attachment to Allah Almighty.

This is the reason why the Cosmos cannot be quantified through scientific rationale. The issue with a secular scientific faculty is to confine the existence of humanity as something insignificant, within an insignificant world, arising from a spontaneous or random origin. This is also defined by science as Abiogenesis, informally 'The Origin of Life' as a process by which life arises from non-living organic matter and eventually evolves into living intellectual beings. They define the process, in vast contradiction to the Qur'an, as molecular self-replication, self-assembly, autocatalysis, and spontaneity of cell membranes, which in summary is a shrugging 'Chance'. The same analysis is applied to the creation of the observable universe, as a series of sequentially accidental events, each one leading to the other like dominoes emerging with human life as it is.

The consequence of this thought process is that the insignificance of humans having evolved as the most intelligent species on the planet, must now strive to advance civilization into the cosmos, or face extinction in similitude to other species before, such as the propagated Neanderthals. This 'Origin of Life' theory is also disposes any religious definition as a 'Creation Myth', defining the creation of the universe as a spontaneous ordering of the cosmos from a state of chaos or amorphousness (formlessness). The ignorant mind then asks foolish questions shameful to human intellect, such as 'Are we alone in the universe?', 'Will we ever travel through the stars and galaxies?' or 'Is Artificial Intelligence the next step in the evolutionary process?' Such foolish minds then look into the night skies in search of 'Alien ships' or will inject their bodies with computer chips with an outward haughtiness of having achieved a great milestone, only adding to the secular notion of man as an insignificant being.

Deprived of profound knowledge, such as of those who are enlightened by the Qur'an, the secular mind strives to discover the cosmos by regarding humanity as 'insignificant'.

In truth, we *are* insignificant in contrast to the

majesty of Allah Almighty's power, but we are also significant by the fact that we are existing as per His Divine Decree.

In his *Diwan,* the fourth Khalifah of Islam, Imam Ali ibn AbiTalib (may Allah be pleased with him) said, *'You think yourself (Nafsaka) as some insignificant thing, and yet within you are all of the cosmos.'* Note that the descriptive word *Nafsaka,* as a derivative of the *Nafs,* denotes an individuality, in the sense that the entirety of the cosmos exists with Knowledge and Sublimity within the 'self and soul', not within the physical body.

The Holy Qur'an describes this as follows;

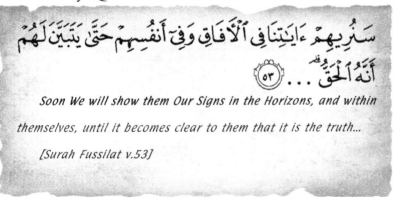

Soon We will show them Our Signs in the Horizons, and within themselves, until it becomes clear to them that it is the truth...
[Surah Fussilat v.53]

The word *'Afaqi'* is the plural form of *'Ufuq'* which is the horizon, or the convergence of the material and immaterial. The description of the meeting place, the horizon, is symbolizing where the material universe (as observable visually) meets with the Heavens, and Allah Almighty then says that *'it will become apparent to us'* within ourselves (*Anfusihim*) denoting an observation or understanding through internal sight.

187

Not with telescopes.

The concept of dimensionality, hence, does not exist in physical confines subject to observation and quantitative study.

Comprehension of cosmology, dimensionality and higher states of existence cannot be acquired through rational analysis. One cannot pinpoint coordinates in the cosmos, because the cosmos does not exist with left and right, east and west, or up and down. By extension, one cannot relate to ascension by way of only physical and biological enhancements.

The purpose behind defining the cosmos as *Samawat* in relation to the root word *Nafs* (verse 53 above), is to clarify and validate that ascension through the cosmos is indeed possible, but only by way of drawing oneself closer to Allah Almighty.

Hence the importance of *Salaat* (prayer) and the emphases of *Dhikr* (remembrance). The virtue and divine importance of *Salaat* is not just ultimately beneficial upon one's existence, but it forms a medium of communication with the Divine Creator, and such a communication, if properly established, is an ascendancy through the Heavens to the Throne.

Allah Almighty tells us in the Holy Qur'an that there will come a Time when transcendences through the Heavens will be made possible for those among righteous believers, albeit in a different context and environ far removed from our current state of biological existence.

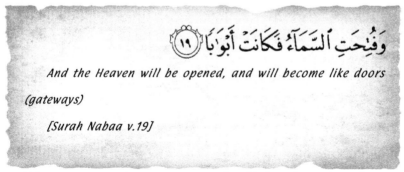

And the Heaven will be opened, and will become like doors (gateways)

[Surah Nabaa v.19]

The verse is denoting, and Allah Almighty knows best, the Day of Judgment when the lowest Heaven, or the Heaven closest to the earth will be opened as a gateway, or doorway, into the Other Worlds.

With relevance to our existence, a study of cosmology does not enable us with the ability to achieve interstellar travel or dimensional travel for the purposes of domination, colonialism or leisure as is widely propagated by the secular minds. This ideology typically arises where fact reaches its limit, curiosity is hardly satisfied, and the mind begins to deviate from a path of true knowledge. A desire to fill the void is then infatuated with a fictitious analogy to traverse into otherworldly existences. Unknowingly, this distortion in the mind and heart is often instigated by Iblees himself, akin to when he attempted to distort the Hearts of Adam and Hawaa by instilling a desire of immortality (*Taha v.120* and *Al-A'araf v.20*).

Oftentimes, in a desperate attempt to actualize such

189

desires, to make them a reality, man is driven to push forward through every boundary, but emerges defeated (*Ar-Rahman v.33*) because sight and perception is veiled, the desire becomes materialistic, and it is at this juncture that Iblees, who first instilled the desire, betrays man to his own spiritual destruction.

There is a dire need for Muslims to comprehend cosmology as is told by Qur'an, and to realize its importance with regards to the purpose of our existence through a proper methodology which is simply an adherence to *Islam, Imaan,* and *Ihsaan.* The underlining matter is that if Allah Almighty has revealed the subject of cosmology to us, then there must be a divinely profound reason. This subject cannot be covered within one chapter, but it serves its purpose in this book as a thought triggering introduction to the next five chapters. We will, *Insha'Allah,* dedicate a separate study to the Seven Heavens in another volume.

THE
VERSE
OF
DARKNESS

For centuries, philosophers and scientists like Plato and Euclid, as well as Ptolemy, argued that the way we see and perceive our environs is by way of emitting light from our eyes onto our surroundings, like a beamed torch. The various theories founded on these arguments, explained that animals saw better in the dark, or at night, because the amount of light their eyes could emit is stronger than human, and this was the basis for applying the word 'light' to the 'light of the eye'.

This was how the philosophical notion of 'the eyes being the windows to the soul' came to be, and it has played a crucial part in modern psychology. By and large, this is the only aspect of the above mentioned theories that still holds a validity and rational explanation, in that, what affects the soul, what upsets the inner 'self', often manifests itself through the body, the eyes being one of the aspects that can reveal an imbalance of any emotional or depressive state otherwise concealed from the outside world. Additionally, within the context, elements of concealment, such as 'lying', can often be betrayed by the eyes, and these elements are as a result of the *Nafs* acting or enacting in a state of desire and sin, known as *Nafs Al-Ammarah* (*Yusuf v.53*).

Going back to the theories of sight being as a result of light emanating from the eyes, this was the basis of all rationality to explain how we, humans, visualized our environs and the purpose of the eyes as an organ of the human body. Chiefly based on these ideologies, generations of physicians diagnosed, or in this case, 'misdiagnosed' any defects and deficiencies in all forms of visual perception.

Of the weak-sighted, they would say 'the light of his eye is weak'.

Of the blurry-eyed, they would say 'the light of his vision is impaired'.

Of the blind, they would say, 'the light of his eye is quenched'.

It was not until the early 11th-century that these ideologies were refuted by Islamic Sciences. In his remarkable seven-volume treatise, *Kitaab Al-Manazir*

(Book of Optics), Hassan Ibn Al-Haytham laid out the foundations for what every scientific breakthrough owes its achievements. He stated, through experimentally founded arguments, that the eye did not function as an emitter, rather as a recipient, by which light would come from the object (as source origin, or reflection) and enter the eye to be imaged by the brain in color, vibrancy, and luminance.

However, all of the above works by Ibn Haytham, as well as others before and after him, only define the function of the eye. They do not explain the essence of 'Light'. They define the processes through which light emanates from a luminous object, how it bounces off a reflective surface, how it bends in refractive mediums, how it enters the eye, and how the image is formed, akin to how the modern camera works, but the question still remains— *What is Light?*

Can we measure Light?

Science will say 'Yes!' presenting us with numerous equations and theories, and some of us will blindly accept that response, yet some of us still shake our heads with uncertainty.

Because the questions then posed, are these;

Can we state, with utmost confidence and irrefutable fact, that 'Light' can be collected and defined in composition and actuality?

Can we 'touch' Light? Can we 'smell' Light? Can we

'taste' it? Can we 'see' or 'hear' Light?

When the sun's rays peer through the window, are we looking at 'Light', or are we merely noticing illuminated particles of dust? When the rays strike the wall across, are we truly seeing 'Light', or merely the defined shadows of the window by which we can deduce a source of Light entering the room?

Let us look at the scientific attempt at rationalizing these thoughts.

As the works of Ibn Haytham were steadily adapted into the infantile foundations of modern science, 'Light' was assumed to be a wave with frequency and amplitude, measurable by way of analyzing oscillations. To a certain extent, some success was viable, until physicists discovered that 'Light' was in fact made up of tiny particles called 'Quanta', which led to the field of science known as Quantum Mechanics. Quantum Mechanics revealed that 'Light' was neither particle nor wave, but *behaved* like particles in some cases and *behaved* like waves in others.

This classified 'Light' as a form of electromagnetic energy, where 'electrically' it performed as particles, and 'magnetically' it performed as waves. In essence, a wave is but as a result of an oscillating or vibrating particle.

This led to the theory that elements which *behaved* in a similar manner were not necessarily 'Light' but also 'Light and Energy' varying in wavelength and particle oscillation. The theory was validated by a series of experimentations through the 1800s, and thus was formed the Electromagnetic Spectrum, with visible 'Light' constituting a tiny portion in the middle.

The question is posed once again; Is this portion of the spectrum truly 'Light', or just a variance of bandwidths both perceptible and not perceptible to the human eye?

By all accounts, if we *can* measure 'Light', we should be able to measure darkness, for what benefit is 'Light' to the eye that cannot see?

You may have noticed then, that we have used the word 'Light' in quotations from the very beginning of the chapter, in similitude of how we used the phrase 'observable universe' throughout this book.

This is because in every fundamental analysis, physics or otherwise, we *cannot quantitatively measure Light*, just as well as we cannot measure Darkness. We cannot measure Cold. We cannot measure Time, Knowledge, Wisdom, Faith, Emotion, these and others exist beyond human rationale, yet well within human comprehension.

We can quantify a *Perception* of Time. We can quantify a *Documentation* of Knowledge. But existing on their subliminal realities, neither of these aspects can be quantified nor perceived in their absolute states of

195

existence or origin.

Similarly, we can measure the *Speed* of Light. We can measure the *Luminosity* of Light, the *Intensity* of Light. But we cannot measure Light itself. The intensity of its visibility, and our perception of it, depends purely on its origin and its destination, and what is perceived is not 'Light', it is the by-product of something essentially luminous.

By this definition, what we perceive as 'Light' is composed of two elements, the source of 'Light' and the effect of 'Light'.

We burn fuel to *emit* 'Light'. The flame of a candle is variant from the flame of a burning log, far variant than burning oil, or the gaseous compositions of distant stars, even the subtle reflection of the moon and nearby planets. No one source of 'Light' is the same as the other.

On the other hand, far from 'Light', we have shadow and darkness, where 'Light' has not yet arrived. Similarly, we cannot measure darkness, except by denoting it as a region or environ without 'Light'.

We can measure the *Attributes* of Darkness. The *Length* of the shadow. The *Duration* of Darkness, but at its core, darkness is simply a definitive term for the benefit of human perception. By its definition, Darkness is just another tool to measure the *Absence* of Light, just as *Cold* defines the Absence of Warmth, as 'nothing' defines the absence of 'everything' or 'anything'.

Consider the dimensionality of the cosmos. Thus far, we have determined that there exists more than the material observable universe in Seven variances as the Holy Qur'an describes. Consider also, that the material observable universe only exists to our perception by the causation of Light. Without Light, we see nothing. We read nothing. We measure nothing. The very essence of our existence is a determinant on Light, just as well as Time and Knowledge. All else is physical and biological.

Now, hold this thought as we progress through the following analysis.

In his *Mishkat-Al-Anwar*, Imam Al-Ghazzali makes a remarkable statement by using very simplistic allegories, and we will be using his study to explain the subject. He writes, *'The Real Light is Allah, and the word 'Light' is otherwise only predicated metaphorically and conveys no real meaning.'*

We urge the reader and researcher to study the works of Imam Al-Ghazzali, and draw from them the immense knowledge contained.

To explain the above statement, we must understand that the word 'Light' is in the employ of three categories of people. The View of the Many. The View of the Few. The View of the Fewest of the Few.

Within the context of the 'Many', Light avails

itself as a phenomenon. Now, a phenomenon, by its characteristic, is a relative term which means *'a fact that is observed in existence, whose cause or origin is in question'*, and *'the object of man's perception, what the senses or the mind realize but cannot explain'*. Thus, a phenomenon is something whose existence and non-existence are relative, both to perception and reality. As 'Light', in this context, its existence and non-existence are relative to perceptive faculties, such as the Senses, and the 'View of the Many' affirm this perception to the most powerful sense, the Sense of Sight. This Sense of Sight has the ability to define three unique instances.

> ***That which has no luminance, and cannot illuminate.*** *Bodies such as the earth, human, animal, objects.*
> ***That which has luminance, but cannot illuminate.*** *Bodies such as the stars.*
> ***That which has luminance, and can illuminate.*** *Bodies such as the Sun, a Lamp, Fire.*

Of the second and third instance is what the View of Many attest the word 'Light', and sometimes to the first category when the expression 'lighted up' is given, such as the 'earth is lighted up', or 'your face is glowing'.

In all three instances, the definition of 'Light' given by the 'View of the Many' is dependent on two things, origin and destination, that is, the source of the Light

198

and the sight of the observer. The View of the Many is also heavily dependent on all things rational, hence the View of the Many is, in fact, the Opinion of the Many, one which heavily disputes and combats the View of the Few.

The View of the Few is as follows. 'Light' is that which appears and causes to appear only to the seeing eye, because it neither appears nor causes to appear to the blind. Yet, despite the lack of sight, perception does not elude the blind. 'Light', therefore, is not observant, nor does observation take place through it, except when it is present. By definition, the 'equation of sight' is incomplete without the element of 'Light'. The View of the Few thus holds that spiritual perception is just as important as visual perception, in fact, more important, in that it observes and also acts as the medium through which observation takes place. By the View of the Few, Sense of Sight exists both externally and internally, giving more importance to internal sight for illumination as opposed to external sight.

Of the Fewest of the Few, *all* perception of Light is internal and spiritual. Their perception, therefore, is like that of Imam Al-Ghazzali, that the word 'Light' has no significance unless it is used to attribute God Almighty as Light.

This view affirms that the perception of visual 'Light' and sight is marked by defect and illusion.

Imam Al-Ghazzali highlights the kinds of defects not as biological or physical impairments, but perceptive shortcomings.

The visual sight can see others, but not itself. It can see forward, but not backward. It cannot see the very distant, nor the very near. It cannot see behind a veil. It cannot see into the future, nor can it see into the past. It sees only the exterior of things, not the interior. The parts, not the whole. The finite, not the infinite. What is large, appears small; what is small, large. What is at rest, appears in motion; what is in motion, at rest.

Like-minded scholars affirm that Light is a perception of sight, and of sight there is more than the physical eye. There is Emotion in the Heart (*Qalb*), Intelligence in the Mind (*Aql*), Thought in the Soul (*Nafs*), and Life in the Spirit (*Ru'h*), each with an eye and a perception of its own, therefore, each with its own kind of 'Light'. The Light of an *Idea* in the Mind is variantly illuminating from the Light of *Love* in the Heart, or the Light of *Thought* in the Soul, or the Light of *Spirituality* in the *Ru'h*.

The reader who has ventured thus far, has understood then that there are *two* kinds of sight— that which is external, and that which is internal. One belongs to the World of Tangible Rationality, the other belongs to the Celestial Realm of Sense and Understanding. Both these worlds have their Light with which 'seeing' is perfected both visually and perceptibly. In one world, we see the bud and the flower, we perceive its colors and its scent, but in the other world, we *feel* its vibrancy, we *appreciate* its beauty as it blossoms, and we *sense* its affectionate arousal. These are all attributes of sight and perception, attributes revealed by Light. Visual sight is revealed by a perceivable imposition of Light, and internal sight is revealed by an intuitive imposition of Light.

What of Darkness though? What of the void where Light is absent?

The Holy Qur'an says;

وَٱلَّذِينَ كَفَرُوٓا۟ أَعْمَٰلُهُمْ كَسَرَابٍۭ بِقِيعَةٍ يَحْسَبُهُ ٱلظَّمْـَٔانُ مَآءً حَتَّىٰٓ إِذَا جَآءَهُۥ لَمْ يَجِدْهُ شَيْـًٔا وَوَجَدَ ٱللَّهَ عِندَهُۥ فَوَفَّىٰهُ حِسَابَهُۥ ۗ وَٱللَّهُ سَرِيعُ ٱلْحِسَابِ ﴿٣٩﴾

But for those who Disbelieve, their deeds are akin to a Mirage in the Desert; the thirsty one supposes it to be water until he comes upon it (the mirage), and finds naught but Allah, He will pay him (the disbeliever) his due; And Allah is swift with settling His dues. [39]

أَوْ كَظُلُمَٰتٍ فِى بَحْرٍ لُّجِّىٍّ يَغْشَىٰهُ مَوْجٌ مِّن فَوْقِهِۦ مَوْجٌ مِّن فَوْقِهِۦ سَحَابٌ ۚ ظُلُمَٰتٌۢ بَعْضُهَا فَوْقَ بَعْضٍ إِذَآ أَخْرَجَ يَدَهُۥ لَمْ يَكَدْ يَرَىٰهَا ۗ وَمَن لَّمْ يَجْعَلِ ٱللَّهُ لَهُۥ نُورًا فَمَا لَهُۥ مِن نُّورٍ ﴿٤٠﴾

Or it (the disbeliever's state) is like the Darkness in a Sea, so deep, A wave covers it, upon it (another) wave, upon it a (thick) cloud; Darkness upon Darkness! When he (the disbeliever) puts out his hand, hardly can he see it; And for whomever Allah has granted no Light, for him there is no Light at all

[Surah Nur v.39-40]

In the *Mishkat,* Imam Al-Ghazzali explains the above verse, which he describes as the 'Darkness Verse', using deeper words, and we will quote his *Tafseer* here;

'But as for the Disbelievers, their deeds are as it were massed Darkness upon some fathomless sea, the which is overwhelmed with billow (wave) topped by billow (wave) topped by cloud: Darkness on Darkness piled! So that when a man putteth forth his hand he well-nigh can see it not. Yea, the man for whom Allah doth not cause Light, no Light at all hath he.'

Veiled is the Heart of the disbeliever, albeit Muslim, Christian, Jewish, Hindu, or Buddhist, if *Nur* has not entered it, or if it has entered but a shade away, it is still veiled, and veiled it is so by the actions of the *Nafs.*

Darkness being a natural state of existence, also becomes so in the Heart of the human who has not allowed Divinity to enter it. Individually and holistically, such are people complementing each other, like the blind leading the blind, deluding themselves of their own existence. They are like he who is in 'the fathomless sea, overwhelmed by wave upon wave upon thick clouds of darkness'.

This 'fathomless sea' is the material world, filled with delusions and adornments (*Kahf v.7*). The cloud is variant spectrum of viscous veils, sublime at the top, thickening towards the bottom. Every lower veil is heavier than the last, all ranging from doubts and opinions, corrupting thoughts, and deviations from truth, dragging the believer into a realm of disbelief, of *Kufr* (covering) from the illumination of the Light of the True Word of Allah, from True Knowledge, as is the purpose of the cloud to blot out the sunlight.

As he is dragged into the upper wave, as is the wave of ferocious attributes, of pride and prejudice, envy and hatred, arrogance and defiance. The final wave drags him into the fathomless depths, from whence he is utterly shrouded of *any* Light. This is the wave of lust, utter submission to desire and temptation, where he *'doeth what he willeth'*, where his Soul is only occupied with sensual pleasures and intoxications, a satisfaction of worldly ambitions, an unquenchable thirst for currency and wealth.

Here, he is among those whose gods are their own images, flashing across screens, each one worshiped based on how much wealth he has accumulated or how charismatically he can entertain others. They consume above consumption. They bore the earth for all its shimmering jewels. They tear the skies with metallic radiation. They fill the air with toxic gas. They plunge the waters with chemicals. They suffocate the forests to construct their own engineered paradises. They replace their children with pets, they confine their elders to homes other than their own. They are the corrupters of the lands, the seas, and the air (*Ar-Rum v.41*). In this modern age, they are the descendants, and the allies of the descendants, of the tribes of *Ya'juj* and *Ma'juj* who were described as *Mufsiduna Fil-Ardh*, spreading corruption in the Land (*Kahf v.94*).

Of such people who have gone deeper than the depths, Allah Almighty says in the Holy Qur'an;

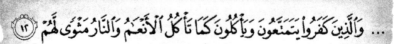

And those who disbelieve, they luxuriate and consume like cattle! For them is the Abyss an abode.

[Surah Muhammad v.12]

All the above mentioned traits are not too far from reach, not even from the Muslim. They are but the Soul's incline away from falling into the depths, no sooner realizing that we end up far from the Light. Darkness upon Darkness, shutting our Hearts from receiving the *Nur* of guidance, wherein our internal sights are veiled from the *Nur,* and external sights are confined to external materialistic vision. How then can such an individual see True Light, True Knowledge, when if he *'puts out his hand, hardly can he see it'?* Then for he whom Allah has not granted Light, has no Light whatsoever.

THE NICHE
THE LAMP
AND THE TREE

As we have identified thus far, a journey through the Cosmos is a journey of the Soul through Seven planes of existence. In lowest of the low, the physical ascension from the physical earth may appear to be a vector displacement to the observer, but to the journeyman, the ascension avails a greater zeal, if but he makes an effort to recognize his journey as opposed to being focused on the destination.

The journey through the Cosmos, is in fact a journey through two elements, both divine in nature. One is a journey through Time, both forward in life, and backward through space. The second is a journey through darkness, from shades of no Light, towards the origin of all Light.

Of the six thousand verses in the Holy Qur'an, the most perplexing, mesmerizing, beautiful, and eloquent verse is to be found in Surah Nur. Across the realm of Islamic Scholarship this verse has been studied, interpreted, and documented in vast volumes and treatises by nearly every prominent Islamic Scholar over the last fourteen centuries since it was revealed. Several notable names such as Al-Ghazzali, Al-Baghdadi, At-Tirmidhi, Al-Razi, At-Tabari, as well as Ibn Hasan Tabarsi, Al-Muhasibi, Ibn Al-Jawzi, and Ibn Arabi, have all made highly commendable efforts in deciphering and drawing the verse's immense knowledge, further simplifying it for our understanding, and we pray to Allah Almighty to bless their efforts and grant them *Jannah*— *Ameen*.

Our humble attempt at studying it within this and the previous chapter is but a speck of dust in the world of Islamic Scholarship, and would require several volumes for us to fully delve into, and explain, its divine meanings.

Regardless, within the context of the study, it is imperative for the reader and researcher to contemplate even a moderate measure of this verse, as it forms the innermost integral structure of protection against the Trials and Tribulations of the modern age, and also provides the believer with an impenetrable shield against the wickedness and grand deception of the Dajjal and his ilk.

It is one of the few verses that require a systematic deconstruction and a block-by-block analysis, as follows;

❊ اللَّهُ نُورُ السَّمَوَاتِ وَالْأَرْضِ

Allah is the Light of the Heavens and the Earth;

مَثَلُ نُورِهِ كَمِشْكَوةٍ فِيهَا مِصْبَاحٌ

An example (allegory) of His Light is a Niche, wherein there is a Lamp;

الْمِصْبَاحُ فِي زُجَاجَةٍ

The Lamp is in a Glass;

الزُّجَاجَةُ كَأَنَّهَا كَوْكَبٌ دُرِّيٌّ يُوقَدُ مِن شَجَرَةٍ مُّبَرَكَةٍ

The Glass as if it were a Star (or a Pearl), brilliant, which is lit from a Tree, blessed;

زَيْتُونَةٍ لَّا شَرْقِيَّةٍ وَلَا غَرْبِيَّةٍ يَكَادُ زَيْتُهَا يُضِيءُ وَلَوْ لَمْ تَمْسَسْهُ نَارٌ

An Olive, not of the east nor the west, whose Oil would glow even if it were not touched by fire;

نُورٌ عَلَى نُورٍ

Light upon Light;

يَهْدِي اللَّهُ لِنُورِهِ مَن يَشَاءُ وَيَضْرِبُ اللَّهُ الْأَمْثَالَ لِلنَّاسِ وَاللَّهُ بِكُلِّ شَيْءٍ عَلِيمٌ ﴿٣٥﴾

Allah guides His Light unto whomever He wills; And Allah sets forth (such) examples (allegories) for (the understanding) of mankind; And Allah is the Knower of All Things.

[Surah Nur v.35]

One must not read, recite, or study the above verse without its eloquent Arabic literacy. Contained within it, we believe, are a *magnitude* of meanings and interpretations, *all* existent (just like the entirety of the Qur'an) without *any* cosmological interference throughout the Seven Heavens and the observable universe (perhaps even beyond). And Allah Almighty, in His infinite Wisdom and Knowledge, is well aware of all things.

Let us first analyze its literal interpretation.

Consider the night sky and our perception of the observable universe, combined with the previous chapter's analysis of Light and the Perception of Light. What we observe as 'Starlight' or 'Moonlight' or even 'Sunlight' is not Light itself, rather a by-product of the object exhausting its fuel and radiating Photons and Quanta of energy which are visually perceptible or detectable.

This energy encompasses the entirety of the Electromagnetic Spectrum from Gama Radiation with the highest frequency of oscillating particles and waves, to ELF (Extremely Low Frequency— Micro-Waves and Radio Waves). Within this spectrum, there is the tolerable frequency which can be detected by the human eye, from Violet (higher frequency) to Red (lower frequency). On either side of this pocket is Ultra-Violet and Infra-Red, just a tad beyond visual detection.

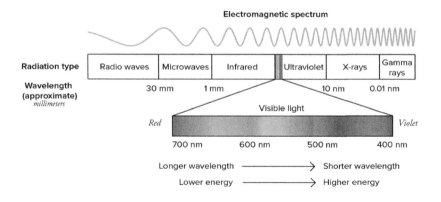

The verse above, also known as the Light Verse, or *Ayat An-Nur,* is clarifying that what we perceive as 'Light' is not actually so, and this is because the essence of 'Light', just like Time, cannot be visually perceived. For this reason, Allah Almighty is using an *allegory* to define Light, one of His attributes as *An-Nur.* It holds, therefore, that Light, like Time and Knowledge, is itself pure, from a source which is pure in itself. From Allah Almighty.

Just as well as God Almighty cannot be explained in definitive and rational terms, so is it difficult to explain His attributes. However, in contrast to God Almighty Himself, His attributes *can* be allegorically explained.

What this means is that, Allah cannot be allegorically explained, because Allah, and His Throne, exist far above, as well as within the entirety of the Worlds, constituting a Realm of Realms where 'up, down, left, and right', have no significance, and without these coordinates, human perception loses all bearing and rationality.

Within the physical universe, some of Allah's attributes

are well observable, identifiable as Signs of His Divine Presence, as they manifest in individual and holistic environs, such as His Mercy as the Most Merciful (*Ar-Raheem*), or His Bounty as the Most Bountiful One (*Al-Kareem*). On the other hand, some of His attributes cannot be realized with visual perception because they do not manifest themselves in tangible forms, such as His Forgiveness as the Great Forgiver (*Al-Ghaffar*), His Ever-Existence as the Eternal Being (*Ad-Dahr*), or His Light as The Prime Light (*Al-Nur*).

The allegorical definitions of these attributes cannot be perceived by external vision, because their manifestations do not occur in tangible forms. Their perception is only possible through *internal sight,* hence the phrase, 'Inspiration from God', and 'the Guiding Light of God'.

The example of Light is defined as something within a niche, a crack, or crevice, which is implicating the great intensity of the Light as it is emitted. Within the crevice is where we find the source of the Light, a Lamp placed inside a Glass, shimmering like the most brilliant of Lights we could possibly find in the night sky. The implication of this statement is that the Light coming out of the crevice, the Lamp, the Glass, and its glow, are, as a source of the Light, *all from the same origin.*

This statement is very crucial to understand, because scientifically, 'Light' can lose its energy through heat, reflection or refraction, but this is a *scientific* analysis of a *perception* of 'Light' which is implying that the strength of the source is always greater than its destination. The Light defined in the verse above, however, is unlike whatever science can define, because it is followed by a very beautiful and profound explanation.

Its comparison to a 'star' or 'pearl' is mesmerizing, in that the Lamp (the source) is inside a Glass. Allegorically, if we take the closest star, the Sun, we find that 'Light' may be radiating from the surface, but its source is *within* the object itself.

The *Nur* of Allah, however, is not starlight. Its emittance is from an source quite unique. Allegorically, the source is defined as a blessed tree, the Olive.

Blessed is the *Zaitun* (olive), both in its medicinal properties and taste, filled with *Barakah* (blessings) and *Shifaa* (well-being). The Holy Prophet (peace and blessings be upon him) said, *'Eat of the Olive and use its Oil, for indeed it a blessed Tree.'* (At-Tirmidhi).

Oil cannot burn without fire or heat, which is a known fact. The phrase *'whose Oil would glow even if it were not touched by fire'* implies not a supernatural ability of this Tree and its Oil, rather the purity of its existence, a source of Divine energy that does not require external flare to 'come alive'. Its relevance is that of an allegoric description of the 'Light' as something which does not abide by any mundane perception of illumination, such as that of electricity to the bulb, or wood for the fire. The adjoining phrase '*not of the east nor west*' can have a multitude of meanings (see below for some examples), but the most astounding metaphoric is that in itself, the Tree (or source of light) is not one of this material universe.

Its illumination does not traverse the vastness of space as one would assume of starlight, rather it takes the form of something which 'reveals' or 'enlightens'. Following the Command of Allah Almighty, it crosses the dimensionalities of the Cosmos, without the need of a carrier, itself acting as a carrier for Time and Knowledge.

An example is of he who is in a physically dark room.

With darkness around, nothing can be seen. As he ignites a candle, that which the candlelight touches reveals itself to his eyes. This is a physical manifestation of a physical 'light' in a physical dimension.

As he sees, he then notices before him the Holy Qur'an, its pages open to be read. He reads and recites, audibly and melodiously. The words touch his heart with emotion, just as they would touch the hearts of any who would hear his voice. Do they *mean* anything, though?

He is reading, or reciting, the words as 'what the Qur'an is saying'. In essence, he is absorbing the visual information through visual sight. As and when he earns his merit, as he who is seeking the *'How'* and *'Why'* for the sake of knowledge, another Light, the True Light, The *Nur,* reveals layers upon layers concealed beneath mere words. They are the meanings and understandings that enter his heart, and with his inner sight he begins to *see*, and the brighter the Light (*Light upon Light*), the more he sees, the more he learns, the more he realizes the absoluteness of True Knowledge.

This Light enters the Heart and illuminates the internal sight of the believer, he who seeks to penetrate the darkness by way of acquiring Knowledge, or Enlightenment. It enters the Heart as Inspiration, Intuition, and Guidance from Allah Almighty, because Light, in and of itself, is an elemental guide, the penetrator of darkness, the ray of hope and warmth, ever soothing in nature and ever welcoming to the human 'self'. This verse is the antonymic connotation of the Verse of Darkness (verse 40), following it in Surah Nur.

Throughout the centuries, Islamic Scholars have derived variant interpretations of this Verse, all harmoniously and equivalently correct in their own respect.

Al-Muhasibi (9th-century) referred to Light as Intellect and Reason in the Mind (*Aql*).

Junayd Al-Baghdadi (9th-century) stated that *'Knowledge of Allah is only possible by a special way of knowing (Ma'rifa) through Divine Enlightenment'.*

According to Hakim At-Tirmidhi (10th-century), *'Allah illuminates the heart by the Light of higher spheres and bestows upon it His Knowledge.'*

Ibn Hasan Tabarsi (12th-century) defines 'Light upon

Light' as *'Prophets who came one after another from one root and continued the path of guidance.'*

Fakhrudin Ar-Razi (9th-century) states that 'Light upon Light' refers to the gathering of light rays and their concentration, as it is said about believers; *'A believer has Four positions. If he receives a favor or blessing, he thanks Allah. If he receives a disaster, he is patient and withstands it. If he says a word, he tells the truth. If he judges, he seeks justice. He is like a living man among the dead when he is among those who are ignorant. He moves among Five Lights. His speech is Light. His deed is Light. His place of arrival is Light. His place of exit is Light. His aim is the Light of Allah in the Day of Hereafter.'*

Al-Qunawi (12th-century) states that *'True Light brings about perception but is not perceived. It is identical with the Essence of the Real (the True and Absolute) in respect of its disengagement from relations and attributions.'*

9th-century Iranian scholar and commentator on the Holy Qur'an, At-Tabari, summarized his *Tafseer* (explanation) and *Ta'aweel* (interpretation) on the verse as follows. (Our comments are in brackets)

"God leads as light to the inhabitants of the Earth and the Heavens (through guidance and inspiration). *He rules the world, which He illuminates during the day, and likewise enlightens the Hearts of Believers. The Niche in the wall is the chest of man* (within which is the Spirit and the Soul in the Heart, like the Lamp inside the Glass); *the*

214

blessed tree stands in the center of the world, in Jerusalem (neither of the east nor the west - the Holy Land where the True Religion of Nabi Ibrahim converges), *like the righteous believer, the oil is Muhammad's revelation* (the Holy Qur'an) *and the 'Light upon Light' is Muhammad's revelation following on Ibrahim's revelations* (all previous revelations between Nabi Ibrahim and Nabi Isa, peace and blessings be upon them)."

All these should not be regarded as the 'opinions' of scholars, thereby riding on the secular assumption that 'every man is entitled to their opinion'. To do so, is to entrap one's mind in the materialistic and secular design of the Dajjal. These are the profound and highly intellectual views of our beloved scholars, all assisting us with a better understanding, by way of which we can take a step closer into receiving the Light from Allah Almighty.

Every righteous man of intellect who delved deep into the oceans of the Qur'an emerged with pearls of varied sizes and colors, each one beauteous and enlightening in its own respect. The verses of the Holy Qur'an are considered to echo through the Seven Heavens in Time and Space, their profound meanings and understandings variant to each Heaven and Earth, neither conflicting the other with any cosmological interference. It is upon every believer to swim and dive into the Holy Qur'an's knowledge and seek out the guiding Lights of the Divine Pearls in its depths.

Ibn Arabi (12th-century) said in his *Al-Futuhat;*

'Were it not for Light, nothing whatsoever would be perceived, neither the known, nor the sensed, nor the imagined. The names of Light are diverse in keeping with the names of the faculties. Smell, taste, imagination, memory, reason, reflection, conceptualization, and everything through which perception takes place are Light. As for the objects of perception, they first possess manifestation to the perceiver, then they are perceived; and manifestation is Light. Hence every known thing has a relationship with the Real, for the Real is Light. It follows that nothing is known but Allah.'

He further said, *'The Real is sheer Light and the impossible is sheer darkness. Darkness never turns into Light, and Light never turns into darkness. The created realm is the Barzakh* (the barrier between life and death; a 'purgatoric realm') *between Light and darkness. In its essence it is qualified neither by darkness nor by Light, since it is the Barzakh and the middle, having a property from each of its two sides. That is why He 'appointed' for man 'two eyes and guided him on the two highways'* (Al-Balad v.8–10), *for man exists between the two paths. Through the inner eye and one path he accepts Light and looks upon it in the measure of his preparedness. Through the outer eye and the other path he looks upon darkness and turns toward it.'*

The relevant wisdom to be drawn from the above profound statements of Ibn Arabi is to fully contemplate the even more profound Hadith of the Holy Prophet when he said *'the Dajjal sees with one eye, and your Lord is not one-eyed.'* (Bukhari and Muslim)

The relevance of the Dajjal and his emergence in the modern age is foremost his personal design and architecture of the modern age to suit his desired requirements and preparedness of global domination. His nature is given by his inability to perceive True Light, hence an inability to perceive intuitive depth, and hence an inability to fill a void within him.

This inability is densely pertinent to his egoistic and arrogant nature, in that he has a desire to dominate, but the desire is not to dominate by force, rather it is to rule by acceptance and recognition. It is a lustful desire to feel wanted and accepted, and it is no different a trait which he continuously tries to impose upon every other human. This is the cancerous attention-seeking desire which has led, to mention but one example, to the modern, 'zombified' addiction to Social Media, which largely presents itself as a platform for the immensely lacking 'attention seeker', detrimentally misused as a means to fill the void within themselves.

However, deep within human nature, such a void can never be filled with materialism, because such a void does not exist on a materialistic realm. Its existence is on a sublime plane within the 'self', the *Nafs*. As is the case with he who has plunged into the fathomless sea, such a void is hence sealed off and left to its doom.

A void is simply defined as a vacuum of 'nothingness', an emptiness which remains so until it is 'filled' with 'something', hence, one must also understand that Darkness is a natural state. Just as Cold is a natural state. Just as Timelessness, Ignorance, Emotionless, Foolishness, and Faithlessness. In all planes of existence and inexistence, these are all natural states of actuality. It is when Knowledge is availed, that ignorance lessens. When Wisdom is availed, foolishness lessens. Just as well when heat prevails, the cold is expelled, and when Light emerges, darkness wanes. Just as well, when God said 'Let there be Light, and there was Light' (*Genesis 1:3*), the void of Darkness, Emptiness, Cold, and Timelessness was removed and replaced with the Cosmos existent before and above us.

Just as the void is a vacuum, by way of an indifference in pressure (basic physics), that which has 'substance' is drawn into that which is void of it. Similarly, the Darkness which is a vacuum void of no Light, draws the Light into it, so long as a passage is opened, a passage like a *Niche in a Wall.*

If we can relate to Divine Light as Guidance, then the Guidance in and of itself becomes Knowledge Embraced, which in and of itself is Illuminating and Enlightening in nature. Here is where we now begin to understand the deeply profound meaning of the verse and its relevance *within* the human 'self'.

218

The ability of the human 'self' to absorb and enact Knowledge can be classified into *Five* categories, beginning with its immediate Sensory ability.

The Sensory ability is self explanatory, as one which relies on external sensory acquisition or actuation, through sight, smell, touch, taste, and sound, and to this sensory ability we can symbolically attest the *Niche in the Wall,* as described by Imam At-Tabari. It is the gap in the human 'self' which allows information (not Knowledge) to enter and exit the 'self'.

Then we have the Imaginative ability, and this is further divided into three stages. Firstly, as the information enters the *Niche,* we process it and imaginatively perceive it into Knowledge in its crude and defined form, as objects with shape and dimension, distance and vector, quantitative and tangible. These are aspects which are themselves opaque in nature. Secondly, as these objects are then refined in the Mind, clarified, disciplined, and controlled, they begin to shed some of their opacities, gradually allowing Light to pass through them. This is the essence of the *Lamp.* Thirdly, the dire adherence of imagination to creativity and innovation is the progressive refinement of this perception, leading to a level of intellect and intelligence, hence an understanding of things in a realm where all these objects of perception are no longer opaque in nature. They now allow Light to pass through them in its entirety. This is attested by the *Glass,* which is opaque in its crude and natural form, becoming transparent through a continuous and rigorous process of refinement and polishing. It allows the *Light of the Lamp* to emanate unaltered and undistorted, and also protects the Lamp from external harm and external influence, such as that of *Fitna* (wickedness) and *Fassad* (corruption). At this level, the believer is far above the Cloud of Darkness, and is able to *see* through any deception with his inner eye (*An-Nur v.36-38*).

Next we have the Intelligent ability, and this is not the ability defined by secular academics as one who has a document to prove it (a Degree or Diploma). Rather it is the level of intelligence that perceives without a deliberation of doubt or dispute within the 'self'. Its Light is unperturbed by *any* influence, external or internal, where internal influence is enacted largely by *Waswasa* (the whispers of *Shaytaan*). This is attested by the *Pearl,* or the *Brilliant Star,* whose Light, protected by the Lamp inside the Glass, is akin to the Light of our Beloved Prophets and the great *Ulamaa* (Scholars) of Islam. They are themselves the *Light-giving Lamps,* through whom Divine Knowledge and Revelation transcends the Heavens to the masses on the Earth.

Now we come to the level of Purity, a level where Prophetic Knowledge is so intricately entwined with Prophetic Wisdom, neither can be distinguished nor separated from the other. It is a level where the Sainthood of Enlightened Scholars thirst, with an *insatiable thirst,* to attain. It is from the roots of this Prophetic Level where the Lamp of Knowledge branches out into every aspect of existence, below and above. It is from these roots that the structure of Religion and Governance is founded on the Earth, branching out into elements of Worship, Submission, Social Elements, Politics, Trade, Academics, and so on and so forth. One stem becomes two branches, two become four, four become eight, and so on, boundlessly. This is the *Tree,* and its attribute is given to a specific Tree denoted by its purity, hence the Divine Purity of all the Prophets of Allah Almighty.

This Tree cannot be found in any random individual, regardless of their level of intellect, their number of doctorates, or the integrity of their secular institutions, which is why this Tree is *'Neither of the East nor of the West'.*

Finally, we come the highest of the highest abilities, which is the Transcendental ability. Its ascension from the Tree of Purity is completely and absolutely Spiritual in nature, transcendental through the Cosmos without hindrance, without difficulty. Again, this ability is possessed by the Holy Prophets of Allah, and those who have abilities Prophetic in nature, such as Al-Khidr (*Surah Kahf*). The Knowledge acquired and emanated from this level is self-luminous in nature, requiring no additional support save from Allah Almighty directly. This is Knowledge which is self-replicating and self-evolving, as Knowledge upon Knowledge. Its level of forbearance and patience is self-illuminating with such an intensity, that in most cases, it is almost visually perceptible by the external observer. It is hence described as the *Oil* from a Pure Tree blessed in nature, *'Whose Oil would glow even if it were not touched by fire!'*

To surmise, that of Sensory perception comes first as the *Niche in the Wall,* where it converts sensory information into comprehensible Knowledge as the *Lamp inside the Niche.* Contained within it is the faculty of Imagination and Innovation as the *Refined and Polished Glass,* which itself is contained within Intelligence as the *Brilliant Pearl or Star,* from which True and Enlightened Knowledge extends every branch of the *Pure and Blessed Tree,* not of the East nor of the West, whose emanation and illumination is Pure and Divine as the *Self Illuminating Oil,* infinitely culminating as *Light upon Light.*

The most corporeal exemplification of this Light, is of that described by Imam At-Tabari, as our Blessed and Beloved, Nabi Muhammad (peace and blessings be eternally upon him).

THE
DIVINITY OF DREAMS

Sleep, in human nature, is immediately understood as the state in which the body attains rest. Categorically, it does not only apply to humans, but to *all* living creatures, and as much as it may be astounding to comprehend, so do plants, and even Jinn. It is in this state of sleep where we perceive and observe, as a naturally ingrained ability, Dreams.

In reality, the state of 'sleep' does not only apply to the daily cycle of laying on the bed, as much as it also includes Unconsciousness, Comatose, Vegetative states, and even Death, all of which denote a state of inactivity. However, even within this *inactivity*, a paramount level of activity endures, such as the brain, the heart, other organs, continue to function, and in the case of death, we know that there is an endurance that every Soul must experience in a state of *Barzakh*, the barrier between the realms of Life and Death (Purgatoric State). The word 'sleep' is therefore less applicable to an inactivity of the being, and more applicable to a State of Existence.

With this understanding, Dreams become less as something perceived or observed, and more as an 'Experience'. In its lowest capacity, the Experience can be affiliated with, or influenced by the Brain, but in its highest capacity, this Experience is not, as widely studied by mainstream sciences, an activity of the Brain, but is in reality an activity of the Soul independent of the Brain.

The purpose of this chapter is not to explain *'What Dreams are'* through any mediums of Psychoanalysis, rather it is to direct our intellectual and intuitive attentions as to *'Why and How Dreams occur'* within Islamic context.

As we have thus understood, secular science has gone so far as to condemn the realism of the Soul and Spirit, confining humanity to physical and biological states of existence, that it cannot explain the reality of Dreams beyond what it classifies as 'Brain activity and imagery'. Given our knowledge of the *Nafs* and *Ru'h,* we can now understand life on higher planes of existence where visual reality is not a sensory act of the body, rather an internal sight by way of Divine Light.

There is an aspect of internal visualization which

constitutes the phenomenons of imagination and creativity, which advance science, to a moderate degree, has been able to map and visualize (but not really understand) by the use of neural imaging. As much as these studies have continually pressed on an ideology that Dreams are exclusively mapped in the brain as neural images, we beg to differ purely on the fact that the Brain cannot see what the Mind and Heart see.

Simply put, the actuality of Dreams occurs on a plane of sublime existence when the body is at rest and the *Nafs* has accompanied the *Ru'h* on a higher plane. What the *Nafs* 'sees' as opposed to what the *Nafs* 'experiences' is a clear distinction between the subconscious relays of the subconscious parts of the brain in the physical realm, denoting the lowest form of Dreams, as opposed to actual experiences on Celestial Realms denoting the higher forms of Dreams.

We recognize that the skeptical self will be opposed to any view expressed herein, because very like the concepts of Time, Knowledge, and Light, Dreams are just as intangible and unquantifiable by the rational faculty.

The reality of Dreams was expressed by our beloved Nabi Muhammad (peace and blessings be upon him) when he said *'When the Time (of the End) draws close, the Dreams of a believer will hardly fail to come true, and true Dreams are one of the forty-six parts of Prophethood.'* (Bukhari)

In another Hadith, the Holy Prophet (peace and blessings be upon him said), *'All that will remain of Prophecy after me will be the True Dream.'* (Muwatta Malik)

As intriguing and perplexing as is the subject of Dreams, a phenomenon widely experienced by every living being on the planet, yet so poorly understood by a vast majority, also holds a crucial place in the modern age, mainly because of the above Hadith. Its supreme importance, as the modern age delves ever so deeper to godlessness, remains the only link of continual Divine Revelation after Nabi Muhammad's final Prophethood. We hence urge the reader and researcher to extend a insightful effort into the study of Dreams and to regard with just as much importance as all other branches of Knowledge.

The scientific study of Dreams is called Oneirology, from Greek *'Oneiron'*, meaning Dream, and the study is chiefly formed around the psychological study of the brain during dreaming, linked with memory formation and mental disorders.

In Islam, however, the study of Dreams is called *'Ilm ul-Ahlam,* and the word *Ahlam* (Dream) is derived from the root word of *Hilm,* which means 'Intellect', denoting a state of Sound Mind (*Aql*). There is a very unique link between Dreams and Intellect, in that both bear roots to a state of confidence, resolution,

226

tranquility, and calmness. Simply put, it is difficult to 'Dream', just as it is difficult to 'Learn', when the Mind is agitated.

A dreamer is hence called a *Haalim*, and what is truly interesting is that someone who has forbearance is called *Haleem*, and one of Allah Almighty's attribute is *Al-Haleem*, the Forbearer, the Clement. Nabi Muhammad (peace and blessings be upon him) was also described as *Al-Ahlama-Naas*, he who has greater forbearance than all of mankind.

What this means is that Dreams are of that state of Mind which is at rest with itself (hence we dream when we are asleep), and the higher the state of tranquility, hence the higher the state of Spirituality, the greater the Dreamer's experience on Celestial Realms, in contrast to Dreams in the Biological Realm, ever subject to physical and mental agitation.

Dreams are classified into three categories.

There are dreams which emanate from the 'self', as memories or enactments of physical experiences in the physical and biological realm. These could either be reenactments of actual events, or influenced by such events, either recent in memory or distant. For the large part, these dreams are hosted by the subconscious mind when the rest of the body is at complete rest, or deep sleep. Oftentimes, the characteristics of Lucid Dreams manifest themselves into this category of Dreams. Lucid

227

Dreams are where a person subconsciously becomes aware that they are dreaming, and are hence able to gain some control and rational decision-making in the dream.

Subconscious Dreams are very vital and valuable in providing a 'private' revelation to the state of one's health and inner well-being (the health of the *Nafs*). They function as windows into the Soul, relaying the actualities of the 'self' which are often overlooked when dealing with the complexities of life, resulting in mental and emotional turmoils, some tolerable, others detrimental. The resultant of overlooking such matters, more so ignoring dreams pertaining such matters, is that he who is then afflicted by states of depression or mental illness often resorts to the use of artificial and chemical drugs which cause more harm than benefit by merely suppressing the symptoms rather than solving the root cause.

It should be noted, while on the subject of psychological affliction, that mental illnesses are rarely the effect of physical or neurological afflictions as much as they are traumatic to the essence of conscious realization, that is, an affliction upon the soul (*Nafs*). The resultant effect of such afflictions is then noticeable on the body, symptoms which cause mental disorders by way of a malfunctioning in neural activity. In their most subtle state they are hardly noticeable, and often controllable, but as the afflictions attain their peak, such as when the trauma is repeatedly inflicted, the effects gush forth from the *Nafs* thereby permanently damaging the physical body and brain. In every single case, chemical inductions into the brain by way of artificial drugs, legal or illicit, are merely symptom concealers. The Divine construction of the human 'self', as is decreed by Allah Almighty, is that it is designed to *heal itself* given the right conditions, such as sleep, meditation (*Dhikr*) and prayer (*Salaat*), all providing the soul with peaceful, tranquil, and spiritual environs. Before symptoms can be realized by the body, they are first and foremost realized by the Soul, whether in real-time or in the subconscious, and one of the most effective ways that the Soul can heal itself is by way of dreams (during sleep) and reflections (during prayer), within which it seeks to restore mental peace and equilibrium.

All praise is unto Allah Almighty who has created these dreams as a medium through which the Soul can not only 'see' itself, but revitalize itself. These dreams also play an important role in consoling the 'self' from the unfulfilled desires of the material world, desires that are deliberately neglected, and rightfully so as they cause more harm than benefit, but can often become obsessions which can coerce the 'self' to deviate from a path of righteousness, or bring harm upon others, directly or indirectly.

If not handled with proper care and attention, such dreams are also prone to heavy distortions and impacts by negative forces, such as the *Shayateen* (the worst and most evil of the Jinn), who have the ability to influence the unprotected *Nafs* (through *Waswasa*)from beyond the Unseen Veil of the Jinn, and this is an endeavor everlasting through every human's life, in both states of wakefulness and sleep.

This brings us to the second category of Dreams, which are 'evil' dreams or 'nightmares', originating from Iblees and his progeny (the *Shayateen*) as a means to enact mischief on the human, and for a large part, as a means to plant an evil seed within the subconscious. A deeper study of this influential trait of the *Shayateen* has been made in a separate book titled *Of Jinn and Man*.

For protection against an evil dream, the Holy Prophet (peace and blessings be upon him) said, *'A good dream that comes true is from Allah, and a bad dream is from Shaytaan, so if anyone of you sees a bad dream, he should seek refuge with Allah from Shaytaan (A'udhu Billahi Mina-shaytaani Rajeem) and should spit on the left (three times), and the bad dream will not harm him.'* (Bukhari)

Lucid Dreaming can also manifest in this category of Dreams, where one finds themselves in control over their own 'self' and is able to 'fight back' the devilish influences. Oftentimes, this becomes a natural instinct for true Believers who are themselves protected from the attacks of the *Shayateen*. This protection, and ability to fight back also comes naturally to the Believer when they recite the appropriate supplications before sleeping. These supplications can be found in the books of *Seerah, Fiqh* and *Hadith*.

The final category is that of Good and True dreams, and these are called *R'uya As-Saliha*. They are Dreams which descend *into* us from Allah Almighty, as part of something being 'revealed'. It is this category that we will explore in this chapter to a greater extent.

Following the previous chapters of the Cosmos and the Divinity of Light, it follows thereof that a greater part of what we 'see' in our existence comes from within. Hence, in addressing the subject of Dreams, we will be dealing with the 'Heart' by probing into the depths of human nature, conduct, and existence, as opposed to the scientific study of neurologically explaining the phenomenon of Dreams.

In his book titled *Dreams,* my teacher, Sheikh Imran Hosein writes, *'Some dreams are divine gifts to the heart, and such gifts come only when the heart is sound, healthy, innocent, and penetrated with faith in Allah Most High. Other dreams are either medicine for the heart, or windows to the heart that allow us to see our own hearts.'*

231

It is imperative to understand the correlation between 'Knowing' and the 'Heart', as it then enables one to understand what the Qur'an defines as Allah's punishment on the disbeliever by sealing their 'Hearts' and 'Hearing' and placing a veil over their 'Eyes' (*Baqarah v.7*), by which they cannot realize nor understand reality beyond material reality. The consequence is that in this modern world, a vast majority of humanity, including some Muslims, cannot realize the age of tribulation upon them. They cannot realize the 'bulging one-eye of the Dajjal', because far removed are they from Imaan by their own inaction to make an effort, and so, far removed are they from realizing the essence and benefit of True Dreams, or even distinguishing them from subconscious imagery and evil dreams.

Most hold the opinion that affiliating oneself to Dreams is an act of *shirk,* which is an ideology that greatly contradicts the *Sunnah* of the Holy Prophet (peace and blessings be upon him). This cancerous school of thought, where Dreams are propagated as fictitious fantasies, is in fact as a result of the impact of the secularization of Thought and Knowledge by a materialistic civilization. It is an ideology that seeks support from science and rationalism and is largely uncomfortable to all matters relating to transcendental experiences, and Dreams are one of the most significant modes of such transcendences.

232

The following verses of the Holy Qur'an deeply explains the relevance and profound benefits of True Dreams to the Believer, both as visionary inspirations, as well as continual Revelation from Allah Almighty, thus enabling the 'seer' to see beyond a 'one-eyed' perception.

أَلَا إِنَّ أَوْلِيَاءَ ٱللَّهِ لَا خَوْفٌ عَلَيْهِمْ وَلَا هُمْ يَحْزَنُونَ ﴿٦٢﴾

Oh! Verily! For the friends of Allah, there will be no fear upon them, nor will they grieve. [62]

ٱلَّذِينَ ءَامَنُوا۟ وَكَانُوا۟ يَتَّقُونَ ﴿٦٣﴾

Those who Believe and are conscious of Allah. [63]

لَهُمُ ٱلْبُشْرَىٰ فِى ٱلْحَيَوٰةِ ٱلدُّنْيَا وَفِى ٱلْأَخِرَةِ لَا تَبْدِيلَ لِكَلِمَٰتِ ٱللَّهِ ذَٰلِكَ هُوَ ٱلْفَوْزُ ٱلْعَظِيمُ ﴿٦٤﴾

For them are the Glad-Tidings (Bushra) in the life of this world, and the Hereafter; No change is there in the Word of Allah; That is the supreme triumph (the highest achievement). [64]

[Surah Yunus v.62-64]

A friendship with Allah Almighty, just like a friendship with anyone else, is built on a foundation of affection. It is not a 'friendship' on the tongue, that is, a 'political friendship'. It is a friendship deeply rooted into the Heart. Through affection, one builds trust and faith, and so long as trust and faith are firmly entrenched in strong foundations, there is no reason to fear, nor despair.

To he who has strengthened his Belief, his *Imaan,* to him are opened the doors of the Cosmos within his 'self', his *Nafs,* and herein does Allah Almighty deliver glad-tidings which come in the form of inspiration, intuition, knowledge, hope, and guidance.

Imam Malik recorded in his *Muwatta, 'Urwah bin Zubair used to say that the words of the Lord: 'They shall receive Bushra in the life of this world as well as in the hereafter...' mean good dreams which a man should himself see or others see for him.'*

At-Tabari recorded that *'It (Bushra) is the good dream that a servant may see or it is seen about him. This dream is one part from forty-six parts of Prophethood.'*

All the constituents of Glad-tidings in the form of Good and True Dreams, which act as continual Revelations, are a Spiritual Phenomenon containing guidances and inspirations which can enable us to *understand* the phenomenons of Prophetic Revelations.

What this means is that, in relation to our modern age and the End Times, most of the Revelation pertaining Islamic Eschatology came to Nabi Muhammad (peace and blessings be upon him) in vague and coded forms of visions and dreams, which we cannot decode for ourselves by merely reading words out of a Book. Dreams require interpretation, and literal translation and explanation is *not* interpretation. The interpretations of the *Prophet's interpretations* of his *own Dreams,* require a certain insight that can only be affirmed and authenticated either by the Prophet himself, or by Allah Almighty through True Dreams, and herein lies the deeper meaning of *Bushra.*

The Holy Prophet (peace and blessings be upon him) said, *'Whoever does not believe in 'Good and True' dreams (R'uya As-Saliha) certainly does not believe in Allah and in the Last Day.'* (Bukhari)

The implication of the above Hadith is grave in its most literal sense, as well as in its interpretive sense, for a Belief in *Aakhira* includes a belief in the *Signs* of *Aakhira,* and a Belief, in *any* of the Signs given to us by Allah Almighty, requires insight and intuitive knowledge deeply rooted in the Heart.

The question then posed, is this;

How can one assure himself of a True Dream from Allah Almighty, and not a mischievous one from the devil?

It begins with one's etiquette and habit. He who is seeking a Divinity with his Lord, must first discipline his own 'self'. He should eat modestly, sleep modestly, talk modestly, and live modestly. He should dislike noise, and embrace tranquility. Most of all, he should strive to acquire a higher state of consciousness and strive to live in the presence of his Lord. Finally, he should enter himself into supplication, and simply *ask* his Lord for a Good Dream.

How would one know? One would just know. It is a *feeling* in the Heart. A pleasantry of overwhelming joy and happiness. A tranquility that cannot be put into words. The Revelation of what Allah Almighty wants to communicate with the True Dreamer becomes apparent in and of itself, and at times may not even require interpretation. He who does not understand must seek an interpretation from a rightfully knowledgeable individual.

For those who would neglect the notion of Good Dreams as direct communication from Allah Almighty, of those who would condemn their actuality, to them we plead; Do not deliberately stray your Heart from a path of Divinity, especially not when Allah Almighty has still kept the door open to receive guidance from Him.

In addition to this, one should not ignore the multitude of Dreams that all the various Prophets of Allah have experienced, many recorded in the Holy Qur'an. These Dreams should be carefully studied and understood, for contained within them was knowledge beneficial to the Prophets, by implication of which they are also beneficial to us.

While some may dispute by saying that 'they were Prophets, and not ordinary people', the invalidity of the claim is such that although they were *Prophets*, they were *human,* who attained the highest level of spirituality, but they *were still human.* Just as we are human, and so not inadmissible to the realm of Good and True dreams, only that we must strive to attain such a level of spirituality as our beloved Prophets in order to distinguish fact from fiction, and truth from falsehood.

The essence of Prophethood upon humanity is as a representation of piety on earth and the Advocacy of Divinity. This advocacy must, in one form or another, transcend a hierarchy, from the Prophets whose lives on earth have passed, to those willing and deserving of spiritual inheritance, enabling us to perform at least a measure of Prophetic Advocacy.

The questions then posed, are these;

By neglecting the essence and importance of Dreams, if we as mundane humans choose not to follow in the footsteps of the Prophets, what then is the direction of our existence?

True Dreams, as one of the forty-six parts of Prophethood, form a medium of communication to Celestial Realms, and so what is the meaning of our existence if we are disconnected from a communication with our Lord?

It holds, therefore, that following a path of divinity does not only hold true to that of Nabi Muhammad who is the Messenger to our Ummah, but to all other Prophets as one of the Pillars of *Imaan*. It, therefore, holds that we must also recognize the validity of Dreams in *their* lives, not only as marvels and miracles, but as actualities, and by extension, we must acknowledge that Dreams form part of the entirety of all Prophetic Sunnah, and therefore part of Islam.

Khalil Gibran once wrote;
In the depths of your hopes and desires,
Lies your silent knowledge of the beyond,
And like seeds dreaming beneath the snow,
Your heart dreams of spring.
Trust the dreams,
For in them is hidden the gateway to eternity.

A Muslim is he who has enabled himself to be granted the doors to absolute knowledge, and the doors to eternal existence in the Hereafter, but the doors do not open themselves. The keys do not fall into his laps. Sight does not rectify itself, nor does the Heart cleanse itself.

To be Muslim, one must take the step forward, and the step after, and so on, until he finds himself upward of the Stairs to Enlightenment, and every step is a sandstone of Qur'anic and Prophetic Revelation.

Of the Muslim who does not strive to enlighten his Heart with a desire to be blessed with a heightened capacity to receive Good and True dreams, is a Muslim who refuses to attain and participate in the last remaining part of Prophethood in the world today.

How then can that Muslim expect to identify and protect himself from the *Fitan* of the Dajjal?

A
VICEGERENT
ON
EARTH

The equilibrium of wonder in any thought process is that the thought is never one sided. When we think of something on an inclined bias, we are naturally thinking of its balancing weight across the scales, oftentimes unknowingly. This how thought processes are formed intellectually, by taking into account all the variables, the good and the bad, the best and worst, the left and the right, the forward and the backward, and finally, the up and the down. Ever since its existence, mankind has always looked deeper and deeper into the night sky, and wondered...

That 'wonder' of the 'above' has always had its equal 'wonder' of the 'below'. Why are we on earth? Why here, and not there, on that twinkling star over yonder?

There is a reason why we have included this particular chapter in the study of Time and Cosmology.

Generation after generation, thousands upon thousands of years of asking that question, and only the Spiritual and Enlightened Minds ever found the right answers. They understood what it meant to be grounded, and to have to crane their necks to look up. They understood what it meant to have to bow down to Him Whose Throne was high above. They understood what the Creator had ordained, His Law, His Doctrine, His Will, As Above, So Below.

Today, in this Modern Age of Science and Technology, this Modern Age which claims, arrogantly and defiantly so, to be exponentially more intelligent and enlightened than every other Age before it, has narrowed down to what it boastfully asserts as the most 'profound' answer it could ever bethink.

Imprisoned on earth.

That is the underlining statement given by every secular scientific mind. We as humans are prisoners on earth due to gravity and a 'not-yet-acquired' ability to venture into the universe.

To a certain extent, this is not only a statement propagated by secular thoughts. Even certain Christianic and Judaic beliefs hold that man was expelled to earth as a punishment for disobeying God, for eating of the Forbidden Fruit.

These crude and ignorant doctrines have somehow even seeped into the minds of Muslims, widely misleading us to believe that we have been deprived of Paradise *because* of a mistake made by the first man and woman, and that the only way we can earn back eternal glory is to be righteous. To a certain extent, they are not wrong, for the only way to earn a place in Paradise (*Jannah*) is by way of righteous conduct.

In satanic and Kabbalistic doctrines, the belief is overwhelmingly variant from mainstream Abrahamic faith. A summary of this demonic doctrine is as follows, and it should be noted that the Kabbalah serves as the foundation for every Dajjalic ideology and its followers.

'But as for the Tree of Knowledge, you must not eat of it, for you shall die' (Genesis 2:17).

Man was prohibited from eating the forbidden fruit, because it was of the Tree of Knowledge. It was Lucifer, the fallen angel, who took a pitying unto man's pitiful ignorance and lack of knowing, and enlightened him to take of the fruit. Upon swallowing the fruit, man's eyes were widened, his brain unlocked. He attained the divine knowledge that was kept from him, and he began to discover the means to Godliness. Fearing this uprise, God exiled man to earth and wiped all knowledge from his brain. On earth, man has progressed through levels of intelligence, superseding all other species to become dominant. It is now, at the pinnacle of the modern age, that man has discovered a means to regain his destiny of being God-like.

(The Kabbalah commentary of Genesis 2:17)

Our focus in this study is not to address these cancerous ideologies of man striving to become invincible and immortal, as a detailed study has been made in another book *The Abyss of the New World*. Here we will look at the more substantive reality of man's existence on earth, and how the secular doctrine of the Dajjal has attempted to poison our minds.

Islam defines man's purpose on earth not to serve as a punishment, but as a test and an advocacy, and to understand it better, we will begin with the creation of man.

Before man, there existed on earth another being, created in its own respect, for its own purpose.

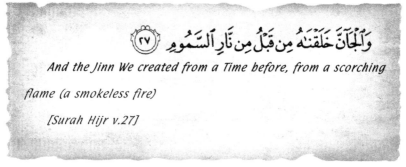

And the Jinn We created from a Time before, from a scorching flame (a smokeless fire)

[Surah Hijr v.27]

There existed on earth, before the arrival of man, another being created of a Smokeless fire, or a scorching flame, both affirming a kind of fire unlike that which we perceive on earth. These are the Jinn.

To quickly surmise their tale, the Jinn populated the earth in large numbers, and as their lives endured, so did their corruptive abilities, and corrupt the earth they did. One among the Jinn was taken up to the Heavens

242

to serve with the Angels, and numerous scholars have differed over the circumstances which we will not discuss here to remain within context. A separate study has been made in *Of Jinn and Man*.

According to Ibn Kathir, Ibn Taymiyyah, and various other scholars, this particular Jinn was named *Azazil* or *Azazael* in Hebrew, and his name was permanently coined to Iblees when he despaired in vain, for *Iblees* means 'he who despaired', derived from the root word *Balasa* meaning 'he despaired' and *Tablis* meaning 'confuse', all building up to his wicked set of characteristics. It is from this root that the Greek word *Diabolos* is derived, meaning Devil.

He was there, in the Heavens, in the company of the Angels, when Allah Almighty made the following declaration;

وَإِذْ قَالَ رَبُّكَ لِلْمَلَـٰئِكَةِ إِنِّي جَاعِلٌ فِى ٱلْأَرْضِ خَلِيفَةً ۖ قَالُوٓاْ أَتَجْعَلُ فِيهَا مَن يُفْسِدُ فِيهَا وَيَسْفِكُ ٱلدِّمَآءَ وَنَحْنُ نُسَبِّحُ بِحَمْدِكَ وَنُقَدِّسُ لَكَ ۖ قَالَ إِنِّي أَعْلَمُ مَا لَا تَعْلَمُونَ ﴿٣٠﴾

And when your Lord said unto the Angels, 'Verily! I will place a Vicegerent on the Earth'. They said (in response) 'will you place in it (the Earth) one who will spread corruption and shed blood, while we glorify you with praise and we sanctify you?' He (Allah) said, 'Verily! I know that which you know not!'

[Surah Baqarah v.30]

The affirmation derived from this verse is that; One, there was already a preceding race on earth, most of it corrupt and wicked. Two, the Angels were not *questioning* Allah's intent, rather they were seeking knowledge from His declaration. Three, Allah Almighty had already intended for man to be on earth as a Vicegerent, an Advocate, and Ambassador, but representing what?

Representing His sovereignty on earth by partaking in His Command and Creation of both the tangible and intangible.

This directly refutes any theory that man was sent to earth 'as a punishment'.

Lastly, despite disclosing one intent of placing man on earth as a vicegerent, Allah Almighty does not, in any way, disclose the rest of His intentions by using the statement *'I know that which you know not'*, which implies that aside from representation, man has another, or *more than one purpose* to fulfill on earth.

As explained, it should be continually noted that even though it is not mentioned in the verses, when this event took place, a particular Jinn was silently observing and listening. It should also be noted, by reading between the lines, that his silence said more than he could possibly conceal, and there is a deeply profound reason to his silence. Hence the statement in verse 33, as we will see below, *'I know what you reveal and what you conceal.'*

The command unto this wholesome company was as such;

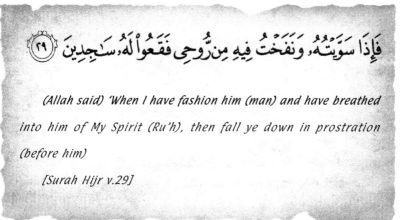

(Allah said) 'When I have fashion him (man) and have breathed into him of My Spirit (Ru'h), then fall ye down in prostration (before him)

[Surah Hijr v.29]

Following what we discussed in the chapter *Of Command and Creation,* one should understand the difference between 'fashioned into body' which is the *Jism,* the 'identity of man' which is the *Nafs,* and the 'breath of life' which is the *Ru'h.* Additionally, the Command to prostrate should not be taken as an act of worship, but as an act of respect to that which Allah Almighty created with a certain uniqueness and honor before the company of Angel and Jinn.

In his treatise, *Al-Bidayah wan-Nihayah,* Ibn Kathir analyzes that the honor of man is on four levels;

Allah's creation of Adam with His own Hand, Breathing into him of His own Spirit, the Command to prostrate before him, and bestowing on him of His own Knowledge (teaching him the names of things, verse 31 below). These four honors give man a superiority, in their own respect over the Angels, who have a superiority of their own respect, and the over Jinn, who have a superiority of their own respect.

Allah Almighty is also ever testing His creation in every instance, as we humans came to realize, beginning with our own creation.

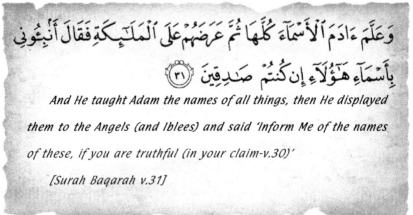

And He taught Adam the names of all things, then He displayed them to the Angels (and Iblees) and said 'Inform Me of the names of these, if you are truthful (in your claim-v.30)'

[Surah Baqarah v.31]

Contrary to popular belief, and all the widespread literal interpretations of the verse, the *'names of all things'* does not imply the literal names in a particular language (the Arabic language, or a language of the Heavens at that Time). One would wonder, and hence lose themselves in speculations over what these objects were, what names were given to them, and in what tongue were they named.

One must take a more a more allegorical, symbolical, and philosophical approach to the phrase *'taught him the names of all things'*, as 'an ability which was bestowed upon Adam', the same ability possessed by the progeny of Adam, which is the ability to identify, define, interpret, and absorb our environs by giving them names and attributes, which in and of itself, constitutes a small, perhaps infinitesimal portion of the Knowledge of Allah Almighty.

What was given unto Adam, hence the pinnacle of Iblees's envy, was an ability not given to the Angels nor the Jinn, and this is the Gift of Knowledge. Where man has the inability of flight and sublimity, like that of Angels or Jinn, man possesses a superiority of intellect and *an ability to gain knowledge* in extended capacities, a trait *not* acquired through the evolution of a species, but acquired as an attribute of the Creator's Design.

This is affirmed in the following verse;

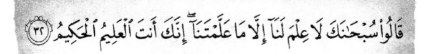

They (the company of Angels) said, 'Glory be unto You! We have no knowledge, save what you have taught us. Verily! You are the All-Knowing, the All-Wise.'
[Surah Baqarah v.32]

What they realized is what they were heedless of, that Allah Almighty had an intent greater than what they knew. By way of 'what they knew' it holds also that Knowledge is not the exclusivity of humankind. The distinction here, and Allah Almighty truly knows best, is that a branch of Knowledge has been bestowed upon mankind, just as other branches of Knowledge have been bestowed upon other creatures.

Note also, that Iblees has been silent all along. He has neither affirmed what the Angels affirmed, nor has spoken out yet. What could possibly have been stewing in his heart, only Allah Almighty knows, but given that he had circled Adam's lifeless body before, floating in and out, deliberating over what this object was and *why* it was being given such importance, one can only imagine his thoughts when Allah Almighty presented them with a test in which only Adam succeeded. Not the Angels, nor he, Iblees, could deliver at the challenge posed before them.

قَالَ يَا آدَمُ أَنبِئْهُم بِأَسْمَآئِهِمْ فَلَمَّا أَنبَأَهُم بِأَسْمَآئِهِمْ قَالَ أَلَمْ أَقُل لَّكُمْ إِنِّي أَعْلَمُ غَيْبَ السَّمَوَاتِ وَالْأَرْضِ وَأَعْلَمُ مَا تُبْدُونَ وَمَا كُنتُمْ تَكْتُمُونَ ۝

He (Allah) said unto Adam, 'Inform them of their names'; When he (Adam) informed them of their names, He (Allah) said, 'Did I not say to you, Verily! I know the Unseen of the Heavens and the Earth, and I know what you reveal and what you conceal.'

[Surah Baqarah v.33]

It should be noted here that Adam (peace be upon him), in the act of 'informing', does not denote a 'point-and-name' synopsis, rather it denotes a complete enactment and thorough articulation of *Understanding, Naming, Defining, Interpreting,* and *Explaining,* all of which comprehensively attribute to the ability to *Absorb Knowledge* and *Deliver Knowledge*. It also follows what the previous verse, 31, denotes Allah Almighty instructing the Angels (and Iblees) to 'inform Him of the objects' as a Command to 'define and explain *what* they are, *why* they are, and *how* they are', to which the Angels responded by saying *'we have no Knowledge save what You have taught us'*, meaning that it was not their inability to identify them, or that they had never seen such objects before, it was that they could not *comprehend* them, hence unable to explain them.

This can allegorically be understood in similitude to the ordinary man who looks up at the night skies and sees the stars. Beyond knowing them as stars, he can neither explain, nor define them, until he has 'learned' them, or until Allah Almighty has bestowed that Knowledge upon him.

Additionally, the word used, *Anba-ahum* (informed them), denotes Adam as the 'informer', and is a derivative of the same root word which also forms *Anbiya* (Prophets), *Nabi* (Prophet), *Nubua* (Prophethood), as he who delivers the Prophecy of Allah (*Nabu'a*), or he who 'informs the people of Allah's message', or 'informs the people of what Knowledge Allah wishes to convey to them' as a directive from Allah Almighty, denoted in the verse as a direct Command from Him as *'O Adam, Inform them of their names'*. The Knowledge thus conveyed from Adam, the Informer, to the company of Angels (and Iblees) was the Knowledge of the objects which they could not name, define, and explain, *and* Allah Almighty's first testimony of *'Knowing what they (the Angels and Iblees) did not'*.

This attributes Adam (peace be upon him) as the 'First Messenger' or 'Prophet' who conveyed to the immediate onlookers a Divine Message from Allah, of which they themselves had no knowledge, but it also elaborates the concept of Man as a Vicegerent, Emissary, and Advocate of Allah Almighty, on earth. It should also be noted that the 'Message' was not only a display of knowledge, or a revelation of the knowledge given to Adam, but also an 'unspoken' message to Iblees who was one of the silent onlookers.

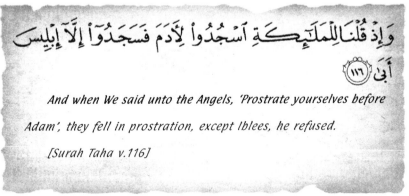

And when We said unto the Angels, 'Prostrate yourselves before Adam', they fell in prostration, except Iblees, he refused.
[Surah Taha v.116]

We chose to continue the narration from Surah Taha instead of Surah Baqarah, for the purpose of continuing within the subject of study, and that is to identify the Purpose of Man and his Relation the to Earth.

Again, as explained above, the prostration was

not in the form of 'worshiping' man, but as an act of respect before Allah's Creation and display of Power and Decree.

Here do we find the manifestation of Iblees's choice between Good and Wickedness, between Obedience and Disobedience, between Acceptance and Defiance, but more so, the resultant manifestation and corruption of his own 'self', likening to he who is falling into the depths and darkness of the fathomless sea (*An-Nur v.40*).

Note, Iblees has not yet spoken at his juncture, nor has he been questioned about his actions.

Note also, that even though it is not explicitly mentioned, it is implied upon the profound thinker and complementer, that there is an element of Time which has endured through and through, from the original intent and *Amr* of Allah, through the Creation of Adam in a lifeless *Jism*, followed by the Spirit of Allah breathed into him, the Time taken for him to learn what Allah Almighty taught him, to the manifestation of the particular event, which is the climactic display of Adam's Knowledge, and Allah Almighty's power. Time has not been mentioned for two main reasons. Firstly, because the events did not occur on earth (as widely assumed by some ideologies pursuing an evolutionary thesis), and therefore earthly Time has no relevance to this event. Secondly, *because* the practicality of earthly

Time is meaningless in the Heavens, the actuality of Time occurring is henceforth incomprehensible to the human mind. And Allah Almighty truly knows best.

فَقُلْنَا يَـٰٓـَٔادَمُ إِنَّ هَـٰذَا عَدُوٌّ لَّكَ وَلِزَوْجِكَ فَلَا يُخْرِجَنَّكُمَا مِنَ ٱلْجَنَّةِ فَتَشْقَىٰ ۝١١٧

We said, 'O Adam, Verily this (Iblees) is an enemy unto you and your wife (Hawaa), so do not let him drive you from paradise, lest you suffer.'

[Surah Taha v.117]

This forewarning, against the harbinger of evil unto mankind, was given well in advance. Here we understand a deeper meaning to several things. One, Adam and Hawaa were *in Paradise.* They were not on Earth, or any Earthly form of paradise. Their *physical* locale was in Paradise. Two, there is an element of 'learning' here that is akin to human nature, in that a forewarning does not necessarily equip one with 'knowing' as much as it contains an element of curiosity, regardless of the gravity of the warning given.

Curiosity drives the curious mind into the bosom of experience, and experience becomes the mother of all lessons. Curiosity is also a plaything of the devil, because curiosity is not rooted into intellect as much as it is into desire. Curiosity is the faculty which lingers between logic and wonder, and it is within its unprotected crevice that the devil seeks to plant his defiling wedge.

Hence the phrase 'Curiosity killed the cat'.

Additionally, there is also an element of faith, akin to that which a parent would expect from their child. When your Lord has forewarned you to steer clear of harm's way, why then would you dare it? Lastly, the forewarning is that of a suffering, but what kind of suffering?

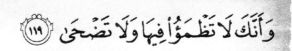

إِنَّ لَكَ أَلَّا تَجُوعَ فِيهَا وَلَا تَعْرَىٰ ﴿١١٨﴾

Indeed, for you therein (in paradise) is no hunger, nor will you be unclothed.

[Surah Taha v.118]

These are the first two kinds of suffering.

وَأَنَّكَ لَا تَظْمَؤُا۟ فِيهَا وَلَا تَضْحَىٰ ﴿١١٩﴾

And indeed, for you therein is no thirst, nor the scorching (of the sun)

[Surah Taha v.119]

Food and Water, Shelter, and Clothing, the basic needs of survival on earth. Elementary science, and a rooted instinct in every human being.

One must now wonder then;

Why do we suffer these needs when Allah Almighty already decreed our being on earth? Why would he forewarn Adam against these sufferings when the earth was already decreed to be our eventual destination?

To answer these questions we have to consult a verse describing a precedence to all these events.

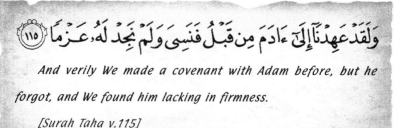

And verily We made a covenant with Adam before, but he forgot, and We found him lacking in firmness.

[Surah Taha v.115]

The word *Qablu* (before), denotes a period of Time *prior* to what transpired. There was an agreement. There was an accord. Could this be the Covenant of Vicegerency between the father of humankind and our Lord, prior to sending us on earth? Rather than indulge in speculations, what we can derive with certainty is that there *was* a covenant between Adam and Allah Almighty, a covenant which he forgot. This means that our purpose on earth had not changed (from destiny to punishment), rather the manner in which the purpose was to be fulfilled had now taken an alternative course, based on something *we* did (meaning 'Adam did').

Herein lies the concept of fate and destiny. Were fate to be described as a 'final destination on earth', destiny would then be describing the process, the 'How we would arrive at that fate', meaning that while fate is sealed by Divine Command, destiny has been let loose for *us* to shape, and Allah Almighty knows best his plans and our thoughts, better than we know ourselves.

Man's destiny is, as it always has been from the moment Iblees became our enemy, resting on a fine thread, balancing ardently between Good and Bad, between Righteousness and Evil, between Paradise and Fire, and it is upon this thread that Man finds himself in the turmoil and battle of a hardened life. It is upon this thread that Man's greatest foe seeks to tip him into the abyss, for to him, why should Man enjoy the fruits of Paradise, while he himself will be burning in the Flame?

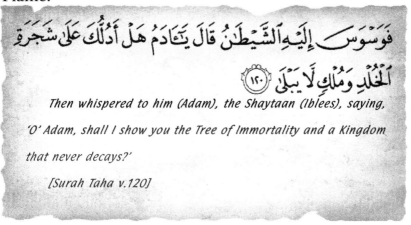

Then whispered to him (Adam), the Shaytaan (Iblees), saying, 'O' Adam, shall I show you the Tree of Immortality and a Kingdom that never decays?'

[Surah Taha v.120]

The word 'Waswasa' literally means 'whisper', but this is not a literal 'whispering in the ear', as much as it is a seduction in thought. Through the veil parting Man from Jinn, Iblees and his ilk do not literally call out to our external senses by directing us toward evil, rather they reach into our 'selves', our Nafs, within which lay the balance between materialism and spirituality. They create the doubt, plant the seed, and watch it flourish into evil deeds as we destroy ourselves without even realizing it.

The devil does not come to us with red skin and horns. He comes disguised as what we most desire in the material universe. He entices. He caresses.

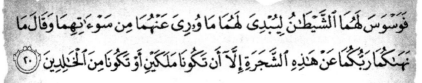

Then he whispered to them, the Shaytaan, to expose what was concealed of their shame, and he said, 'Did not your Lord forbid you of this Tree, only that you would become Angels, or you would become of the Immortals.'

[Surah Al-A'araf v.20]

He cantillates the hymns most prickling to your curiosity.

He instills the melodies most soothing to your soul.

He plays the symphonies most mellifluous to your heart.

He steals your confidence.

He embezzles your trust.

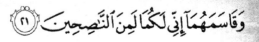

And he swore to them both, 'Verily, I am to you among the (most) sincere advisers.'

[Surah Al-A'araf v.21]

And he betrays you at the very edge, at the very pinnacle, at the very last moment of direness, leaving you to your own discomfiture at the fault of your own self.

فَدَلَّـٰهُمَا بِغُرُورٍ فَلَمَّا ذَاقَا الشَّجَرَةَ بَدَتْ لَهُمَا سَوْءَ ـٰتُهُمَا وَطَفِقَا يَخْصِفَانِ عَلَيْهِمَا مِن وَرَقِ الْجَنَّةِ وَنَادَنهُمَا رَبُّهُمَا أَلَمْ أَنْهَكُمَا عَن تِلْكُمَا الشَّجَرَةِ وَأَقُل لَّكُمَا إِنَّ الشَّيْطَنَ لَكُمَا عَدُوٌّ مُّبِينٌ ﴿٢٢﴾

So he lured them by deception; then when they both tasted of the Tree, exposed to them was their shame, and they concealed themselves of the leaves of Paradise; And their Lord called unto them, 'Did I not forbid you from this Tree? And I told you, verily Shaytaan is an open enemy to you.'

[Surah Al-A'araf v.22]

Despite the ominous warning from Allah Almighty, they were indeed enticed, and they did indeed commit an act of disobedience. Not that they were weak-hearted, but that they were less conscious of Iblees and his malice.

Did they learn their lesson though?

قَالَا رَبَّنَا ظَلَمْنَا أَنفُسَنَا وَإِن لَّمْ تَغْفِرْ لَنَا وَتَرْحَمْنَا لَنَكُونَنَّ مِنَ الْخَـٰسِرِينَ ﴿٢٣﴾

They called (unto Allah), 'Our Lord, we have wronged ourselves. If you do not forgive us and have mercy on us, surely we will be among those lost.'

[Surah Al-A'araf v.23]

A mistake was acknowledged.

Atonement was sought.

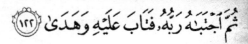

ثُمَّ ٱجْتَبَٰهُ رَبُّهُۥ فَتَابَ عَلَيْهِ وَهَدَىٰ ﴿١٢٢﴾

Then his Lord chose him (Adam), and turned to him (with

forgiveness and mercy), and guided him.

[Surah Taha v.122]

Here lies a conundrum to those who are lamenting man's purpose on earth. Allah Almighty *forgave* Adam and Hawaa. He showed them *mercy* (from His affection). He *Guided* them.

Where is the punishment, if the crime was forgiven? There is not a *single verse* in the Qur'an which speaks of life as a punishment on earth.

قَالَ ٱهْبِطَا مِنْهَا جَمِيعًا بَعْضُكُمْ لِبَعْضٍ عَدُوٌّ فَإِمَّا يَأْتِيَنَّكُم مِّنِّى هُدًى فَمَنِ ٱتَّبَعَ هُدَايَ فَلَا يَضِلُّ وَلَا يَشْقَىٰ ﴿١٢٣﴾

He (Allah) said, 'Descend from it (paradise) all of you (Adam,

Hawaa, and Iblees), some of you (of mankind) unto others of them

(the worst of the Jinn) sworn enemies (of each other). Then when

comes to you, from Me, guidance (messengers with revelation),

then whosoever follows My Guidance will not go astray nor

suffer.'

[Surah Taha v.123]

Thus was our placement on earth, perhaps unlike what was originally planned, but its purpose no less defined. A purpose clearly outlined in the Holy Qur'an, which requires no further interpretation nor explanation, other than that which is declared.

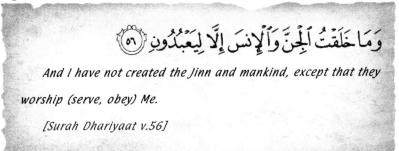

وَمَا خَلَقْتُ ٱلْجِنَّ وَٱلْإِنسَ إِلَّا لِيَعْبُدُونِ ۵۶

And I have not created the Jinn and mankind, except that they worship (serve, obey) Me.

[Surah Dhariyaat v.56]

Iblees's behavior validates that the Jinn are not flawless.

Adam and Hawaa's mistakes validate that Mankind is not flawless.

We all err.

We all need guidance.

The superior ability of the Jinn to transcend matter in Time and Space enables them to survive in their own defined world and state of existence.

Man's superior ability in intellect and knowledge enables Man to survive in his own defined world and state of existence.

Man has a third ability, given unto him through the Breath of Life from his Creator, an ability which draws him, should he choose to embrace it, towards the Heavens.

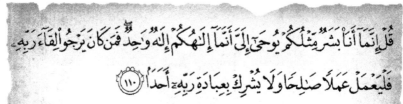

قُلْ إِنَّمَا أَنَا بَشَرٌ مِّثْلُكُمْ يُوحَى إِلَيَّ أَنَّمَا إِلَهُكُمْ إِلَهٌ وَاحِدٌ فَمَن كَانَ يَرْجُواْ لِقَاءَ رَبِّهِ فَلْيَعْمَلْ عَمَلاً صَالِحًا وَلَا يُشْرِكْ بِعِبَادَةِ رَبِّهِ أَحَدًا ﴿١١٠﴾

Say (unto your fellow man), 'I am only a man, like you; It has been revealed to me that your God is only One God. So whosoever seeks to meet his Lord, let him do righteous deeds, and not associate in the worship of his Lord, any other.'

[Surah Kahf v.110]

The last verse in Surah Kahf, the chapter in the Holy Qur'an anchored at the heart of the Signs and Knowledge of End Times, defines man's one and only purpose if he is sincere to his nature, and that is to return to his maker.

يَأَيُّهَا الإِنسَانُ إِنَّكَ كَادِحٌ إِلَى رَبِّكَ كَدْحًا فَمُلَاقِيهِ ﴿٦﴾

O ye of mankind, verily you are toiling onward to your Lord, painfully toiling; (surely) you will meet Him.

[Surah Al-Inshiqaaq v.6]

The secular 'self', hence the one-eyed 'self', only sees life as a hardship, as a punishment, as an imprisonment. He sees the earth as harshly binding, laboring and grinding, and ever does he strive, not to pull through the test and succeed, but to make himself and his life as comfortable as he possibly could.

The believer, hence the 'self' who is able to see with *both* eyes, sees life in a different and more accurate perspective.

Our toiling on earth *defines* our *purpose* on earth. It defines our servitude. Our firmness and resolution. It shapes us, strengthens us, and enlightens us. It gives us the ability to *earn* our way and our place in the Hereafter, in the company of our Creator.

The believer sees, with his external sight, the marvel of creation, and he glorifies Allah Almighty in everything. He sees, with his inner sight, the Guiding Light, Beauty beyond Death, Eternity beyond the Hereafter. He who strives to follow that Light, verily to him is unveiled a glimpse of True Paradise beyond, and his nostrils are caressed with its invigorating fragrance. To the believer, there is naught else but love and affection in his Heart.

This bond of love and affection between Creator and Creation exists throughout all that which exists as a creation. Sadly so, that the one-eyed observer cannot *see,* cannot *realize,* cannot *feel* what he should *feel* as a human being. Sadly so, that the one-eyed observer is enticed not with the Eternity of the Hereafter, in the presence of Allah Almighty, but in the material desire of this world.

Heartbroken and melancholy we are, to see, with our own eyes, that many among them are our own brothers and sisters in Islam. Muslims. Muslims who were born Muslims. Muslims who are clouded with materialism and modernization. Muslims who have taken sectarian and opinionated pledges to be their guiding principles. Muslims who hardly adhere to the *Fardh* foundations of this Glorious Deen, let alone the Sound *Sunnah* of our Beloved Messenger (peace and blessings be upon him).

We pray for them, that Allah Almighty have mercy on their Souls, that Allah Almighty may bestow them with the inspiration to turn back to their Lord. We pray for them to realize their ways, to rectify their ways. We pray for them, from deep within, with tears in our eyes, and tears in our Hearts.

We pray for them.

We pray for them.

We pray for them.

THE
LONESOME ROAD

Islamic Eschatology is a subject so dense and so vast, one cannot simply dive in and expect to unravel the entire mystery in one swipe. For the greater part of understanding it as a Science of Interpretation (*Ta'weel*) one must regard it as a wholesome study of life itself and all the elements surrounding life. As we continue to write these series, the reader will often find continual references to Time, Light, Dreams, Cosmology, and Dimensionality, hence the reason why we have introduced the Islamic Sciences of Chronology and Cosmology.

In a vile attempt to quash the Believer's inner eye, the secular world, through decades of mainstream broadcasting, has largely succeeded in clouding the minds of the populous into believing that Sciences are the subjects of 'Nerds' and

'Geeks'. That Sciences hardly have the ability to generate notes and coins, and therefore less significant to pursue in life. In the Muslim world, nine out of ten Muslims may not even have *heard* of Islamic Sciences, and even Muslims studying Secular Sciences in Secular institutions are unable to find harmony within themselves, because we as an *Ummah,* have forgotten our own curriculum.

We have forgotten the integrity of Islamic Knowledge. We have forgotten the capacity of *'Ilm.* We have capped the immensity of *'Ilm,* confining it only to religious doctrines, forgetting that Sciences also constitute parts of our existence, bearing just as equal, if not less, a knowledgeable importance as religion.

We have forgotten that Knowledge, Earthly and Heavenly, external and internal, plays just as equal a role of bringing peace, stability, tranquility, and spirituality into the Heart and Mind of every believer and to society as a whole.

We have forgotten the great scientists and scholars of our own faith. Men and women who lived with integrity and studied with integrity. Men and women who were armed with knowledge, who could not be shaken, not by the devils of Mankind, nor by the devils of Jinnkind. Men and women who upheld themselves with the honors of knowledge and intellect, not with how tall their buildings were.

Sciences are not just *Lab Studies* of Physics, Chemistry,

and Biology. Sciences incorporate every mundane aspect of our lives, Sociology, Psychology, Philosophy, Arts, have all been secularized by the modern age, leaving very little room for the Muslim mind to find any affection in his Heart to study them from their roots in Islam. In the end, many are deceived, many leave the fold of Islam, and those who remain would only pursue it for the sake of marvel, thus succumbing, often unknowingly, to an urge to satisfy a mere entertaining need.

While a study for the sake of pure marvel and glorification of Allah's Creation is also vital and crucial to the 'self', we must also assert the Islamic Philosophical and Spiritual studies of the sciences of Time and Cosmology for the sake of knowledge, chiefly because the Qur'an is bringing these concepts to our attention, and chiefly because it holds a vast majority of clues to decipher our modern lives. Without the ability to find and study sciences within the fold of Islam, one lacks the ability to see beyond a subjective veil, and the subjective veil is always morphing and shaping itself to keep the mind imprisoned in a material world.

The complexity by which we live today has never before been heard of in our history. Human population, along with every one of its material and societal needs and requirements have all grown exponentially so, clinging on the cusp of utter chaos and disorder.

Try as much as we might to deny it, we alone, a

generation of the last few decades, bear a crude complexity in society that humanity has never before experienced, and it is because we *are* living in *the* Age of Trials and Tribulations. Our food has lost its natural taste and nutrient, our clothing has become transparently shameless, our shelters are burdens upon the earth and every other living organism on it, that we have absolutely lost and forgotten our purpose of Vicegerency on earth.

Despite the vastness of Knowledge available to us, through both Religion and Science, a majority of us choose to be entwined in the here and now, choose to believe in a utopia governed by technology and information, while turning a blind eye from the irrefutable fact that technology and information are, by design, purposed to keep our Minds and Hearts confined.

It is due to this confinement that we fail to realize a very crucial prophecy made by Nabi Muhammad (peace and blessings be upon him) with relevance to Time, its relativity and exponential motion, when he said, *'The hour shall not be established until Time is constricted, and the year is like a month, and a month is like the week, and the week is like the day, and the day is like the hour, and the hour is like the flare of the fire.'* (At-Tirmidhi)

We are living in a society whereby it is becoming more and more difficult to detach from the device than it is to acquire one. Steadily and surely, we are losing what is left of our humanity, becoming mere shadows of ourselves.

We are less a presence in humanity than we are images and usernames on screens, ever so grateful for the 'like' and 'comment', than a visit from a relative, or a gathering whence knowledge and righteousness can be elevated. We are so enticed by posing for a 'selfie' or taking a picture of our food before we eat it, than we are of beautifying our inner selves, or thanking God Almighty for the food He has given us.

What has become of our lives but a series of clicks and swipes, and the urge to fill the void with a device, instead of just gazing out to the stars and marveling at the beauty of creation?

Is this the Vicegerent of Allah Almighty on earth?

Is this our destiny?

Of we who sit with our arms and legs crossed, awaiting our savior, Muhammad bin Abdullah Al-Mahdi. Sure he will come, as sure as the night and the day. Sure he will restore the Khilafah and bring the Muslim world back to its glory. But of we who are still seated with our arms and legs crossed, awaiting him— why should he save us? We who do not even make an effort to elevate ourselves with knowledge and enlightenment. We who are content with delving and pleasing ourselves with the shimmering enticements of the Dajjal, corrupting our own Hearts and Minds— why should *anyone* save us, if we cannot make the effort to save ourselves?

The godless world will do as it pleases. *Ya'juj* and *Ma'juj* will do as they please. The Dajjal will do as he pleases. All will suffer the consequences of their actions, but *we* who have the ultimate weapon of all, the Knowledge and Divinity of the Qur'an, have to do what we must. For all the suffering endured by the Muslim world today, is not because *they* are more powerful than Allah's Might, but because *we* have crippled ourselves, by lowering our integrity as those who have the best ability to acquire Knowledge, yet we fail to make the effort. As those who once lived with an integrity and dignity akin to our Prophets and all their Companions. Our suffering today is all but by our own hands. The Holy Prophet (peace and blessings be upon him) *said* that we will be destroyed *'when immorality prevails'* (Bukhari)

Can we ignore the words of the Holy Prophet, or will we attempt to 'interpret' them in a manner that only suits us and our conveniences, without any knowledge of what is being spoken or why it is being spoken?

The methodologies of acquiring Knowledge have not been invented or reinvented. They have existed for eons, every generation, every nation, taught by a Teacher, a Messenger, a Leader, and a Prophet, all adhering to Knowledge revealed from the *A'arsh* of Allah Almighty.

It is the modern age, and the tyrants of this modern age, who have invented and designed a new method of 'learning', teaching naught but habituating the human mind into their preferred states of existence. When we speak of *Ya'juj* and *Ma'juj,* their infiltrated methodologies into the world of Islam have caused our minds to picture mythological creatures, disguising their realities as humans who have taken to the global scale of domination. When we speak of the Dajjal, his infiltrated methodologies into the word of Islam have caused our minds to picture a supernatural being, disguising his own reality as a simple human with a wicked ambition.

They have taken over our Minds and Hearts, confined what should be Knowledge into mundane forms of information, morphed what should be intellectual and enlightened views into mere opinions, so that while their ambitions endure, we are left groping in the dark. So that when they shine their artificial torches, we are drawn to them like moths to a fire. We were warned of them, and we were warned of what would happen should we join them.

> يَٰٓأَيُّهَا ٱلَّذِينَ ءَامَنُواْ لَا تَتَّخِذُواْ ٱلۡيَهُودَ وَٱلنَّصَٰرَىٰٓ أَوۡلِيَآءَ بَعۡضُهُمۡ أَوۡلِيَآءُ بَعۡضٖ وَمَن يَتَوَلَّهُم مِّنكُمۡ فَإِنَّهُۥ مِنۡهُمۡ إِنَّ ٱللَّهَ لَا يَهۡدِي ٱلۡقَوۡمَ ٱلظَّٰلِمِينَ ۝

O ye who Believes! Do not take (certain) Jews and (certain) Christians as your allies, those who are allies of each other; and whosoever among you takes them as allies, then indeed he is one of them; Verily! Allah does not guide the wrongdoers.

[Surah Ma'idah v.51]

Not *all* Jews. Not *all* Christians. Those among them who are allies of each other. Why should Jews and Christians

269

befriend each other? One says "you killed our 'god'!" The other says, "you are worshiping a false messiah!" (*Baqarah v.113*)

So why should they be allies of one another?

What is their allegiance based on?

What are their ambitions?

Who is their leader?

And why should *we* be the ones to moderate and compromise *our* principles for the sake of integrating with *their* modernizing way of life?

The questions we then need to ask ourselves are these;

Where are we now?

Where are we going?

Where are *our* allegiances?

From shirt, pant, and cellphone, to house, lifestyle, and career, are we *their* allies, or are we noble servants of Allah Almighty and His Deen?

Is it enough to merely go through the physical motions of Islam, praying five times a day, fasting, paying *Zakaat* and performing *Hajj?*

Is there not more to our Religion, our culture, our integrity, and our nobility as Muslims?

There are only two paths to take for every human on the planet.

The path of Darkness or Light.

The path of Good or Evil.

One path takes us away from the comforts and luxuries of this world of the modern age, and towards a path of sacrifice, knowledge, and spirituality, leading to the Lamp of Light upon Light.

The other path tempts us with the modern age's conveniences and material pleasures, but plunges us below the cloud, below the wave, and the wave, and into the fathomless sea of Darkness upon Darkness.

There is no 'in-between'. There is no compromise or moderation. There is no such thing as Liberal Islam and Moderate Islam. Islam is Islam. All else, is all else.

The choice is ours.

If the message from this book, and many others of the like, have been successfully conveyed to the 'seeker' and the 'seer', then to him a door has been opened into a realm beyond the material. A celestial realm of Knowledge and Enlightenment. A realm where the Muslim is not just, but a Believer from his Heart. He uplifts himself to the integrity and dignity of the likes of all of Allah's Prophets and all their Companions, and all those who are, in Allah's Eyes, favored and blessed.

To paraphrase the poetic words of Allama Muhammad Iqbal;

Elevate your 'self' to such a crest, that before it is written, God Himself asks you 'What should I write of your Destiny?'

He is successful, who has fully acknowledged and embrace the *two* sights of man, external and internal, and is now prepared to step through the door in search for the convergence between the two oceans of knowledge. These two oceans will carry the 'self' to an elevation of enlightenment and into the Realm Celestial. In this celestial realm are worlds of marvel, remarkably superlative, to which this mundane world of biological and external sight wanes with limitation.

He who cannot, or will not acknowledge that Light Divine is from the Qur'an, and that the Cosmos resides within the *Nafs* of every Believer, and that only at the convergence, at the horizon (*Fussilat v.53*) can he truly *see*, has been defined in the Book of Allah as;

وَلَقَدْ ذَرَأْنَا لِجَهَنَّمَ كَثِيرًا مِّنَ الْجِنِّ وَالْإِنسِ لَهُمْ قُلُوبٌ لَّا يَفْقَهُونَ بِهَا وَلَهُمْ أَعْيُنٌ لَّا يُبْصِرُونَ بِهَا وَلَهُمْ ءَاذَانٌ لَّا يَسْمَعُونَ بِهَا أُوْلَٰئِكَ كَالْأَنْعَٰمِ بَلْ هُمْ أَضَلُّ أُوْلَٰئِكَ هُمُ الْغَٰفِلُونَ ﴿١٧٩﴾

And certainly We have created Hell for many among the Jinn and Men. They have Hearts but they cannot understand; They have eyes but they cannot see; They have ears but they cannot hear; They are like cattle— Nay! They are more astray, for they are neglectful.
[Surah Al-A'araf v.179]

Thus it was for man who reduced himself to an earthly domain, of flesh and bone. Thus it was for man who was born without the 'knowing', save for the infant's innocence and purity. Thus it was for the child to grow in learning, of life and its demands. Thus it was, and is, and will be until the end, that upon a lonesome road, the pilgrim of life must strive for ascension, to approach a plane of Presence Divine in the Realms of the Highest of High.

Would that man ignore the calling, the Message? Would that man subject his thinking and perception to a material world? Would that man subject himself to the daunting symphonies of the Dajjal as he blows the Piper's Pipes? Would that man become of the likeness of the rats who follow the Impostor?

Are we human… or cattle?

FROM THE AUTHOR

I spent most of my early teenage years engrossed in music and arts. Forgotten was my childhood when I excelled in the recitation of the Holy Qur'an, fluent in the Arabic language, versed in Hadith, Fiqh, and Seerah. Everything forgotten in time.

So obsessed was I with achieving the societal status quo of popularity, that I was on the verge of forgetting my Islam entirely. But, no matter how enticing the music was, no matter how loudly the crowd cheered on every performance, something deep inside always felt missing.

To fill the void, I took to my creative side and began to write, pouring my thoughts and feelings into poetry and stories. The deeper I delved, the more I thirsted, but when the world yielded little else but false hope, I was torn. Torn between making something of myself, and what my religion preached.

Everything I learned about Islam only told me what I was supposed to do. Scarcely did I find the answers I sought... Why? How?

So driven was I, as the reality of the world unfolded itself, I took it upon myself to find my answers. Conspiracies, Geo-politics, Injustice. Why could Islam not provide an explanation for what was going on?

It was because I was asking the wrong questions.

Until I understood that *only* the Holy Qur'an could explain all things, and the Holy Qur'an was sent to a people who could think. So I began to think. Studying and restudying everything I knew and thought I knew, seeking answers from none other than my Creator.

It has taken me years to shed my former self and give my life a purpose of Knowledge, and I have used the gift of words, bestowed upon me by my Creator, to share my thoughts with others.

I am not a conspiracy theorist. I am not looking to start a cult, to create a sect, or to spread my own personal doctrines. I have no personal doctrines or opinions. I am only looking to share what I have learned with like minded individuals, individuals who are also seeking the same answers as I.

To you then, I reach a welcoming hand.

 abubilaal@ironheartpublishing.com

 @authoryakub

 @abubilaalyakub

 @abubilaalyakub

MORE BOOKS BY ABUBILAAL YAKUB

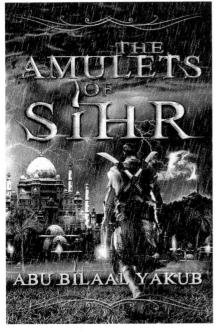

THE AMULETS OF SIHR

Part One of a Four-Part Fantasy Series depicting an Islamic World with essential morals, virtues, and culture.

Mukhtar is a young blacksmith, facing everyday struggles to support himself and his widowed mother.

Life is brutal and harsh, even harsher while the empire only looks after its own, and the rest of the people are left to fend for themselves. In an impulsive moment Mukhtar frees four slaves from their captors. Little does he know how this would shape his destiny. As the turmoil unfolds, his mother unveils her most guarded secret - an ancient and powerful amulet once belonging to his long-lost father. The Amulet sets Mukhtar on a path to unraveling a grim and dark part of his bloodline.

Now, at the crossroads of good and evil, he must face his life's greatest trials in order to save the empire from annihilation.

Enter the realm of the Unseen...

Prepare to face the evil beyond the veil...

THE
THREE
QUESTIONS

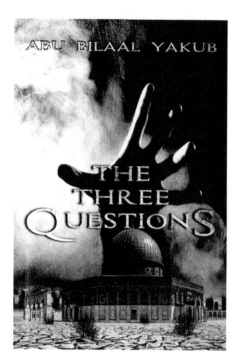

Part One of a series titled *The Impostor and The Two Tribes,* focusing on a unique unfolding during the Holy Prophet's time, which resulted in the Revelation of the anchor of Islamic Eschatology in the Holy Qur'an

Close to the end of the Third Meccan Period, between 619 and 622 AD, in a desperate attempt to foil the unstoppable spread of Islam, the Ruling Tribe of the Qur'aysh sent a delegation to the Rabbis of Yathrib, returning with Three pivotal Questions to test the Holy Prophet of God.

Three Questions that have sculpted the fate of mankind into the modern, secular age we live in today. This book explores these three questions to pierce the godless veils of deception, and better understand the strange unfolding of event in the world, hellbent on ushering the harbinger of evil, the Impostor Messiah, and the dawn of the End of Times.

Printed in Great Britain
by Amazon

81405349R00161